THE UNITED NATIONS UNIVERSITY/THIRD WORLD FORUM

STUDIES IN AFRICAN POLITICAL ECONOMY

The Crisis in African Agriculture

THE UNITED NATIONS UNIVERSITY/ THIRD WORLD FORUM

STUDIES IN AFRICAN POLITICAL ECONOMY

General Editor: Samir Amin

The United Nations University's Project on Transnationalization or Nation-Building in Africa (1982–1986) was undertaken by a network of African scholars under the co-ordination of Samir Amin. The purpose of the Project was to study the possibilities of and constraints on national autocentric development of African countries in the context of the world-system into which they have been integrated. Since the 1970s the world-system has been in a crisis of a severity and complexity unprecedented since the end of the Second World War; the Project examines the impact of this contemporary crisis on the political, economic and cultural situation of Africa today. Focusing on the complex relationship between transnationalization (namely, the dynamics of the world-system) and nation-building, which is seen as a precondition for national development, the Project explores a wide range of problems besetting Africa today and outlines possible alternatives to the prevailing development models which have proved to be inadequate.

TITLES IN THIS SERIES

M. L. Gakou
The Crisis in African Agriculture
1987

Peter Anyang' Nyong'o (editor)
Popular Struggles for Democracy in Africa
1987

Samir Amin, Derrick Chitala, Ibbo Mandaza (editors)
SADCC: Prospects for Disengagement and Development in Southern Africa
1987

Faysal Yachir
The World Steel Industry: Dynamics of Decline
1987

Other titles in preparation.

THE UNITED NATIONS UNIVERSITY/THIRD WORLD FORUM

STUDIES IN AFRICAN POLITICAL ECONOMY

General Editor: Samir Amin

The Crisis in African Agriculture

Mohamed Lamine Gakou

The United Nations University

Zed Books Ltd.

London and New Jersey

The Crisis in African Agriculture was first published in 1987
by:
Zed Books Ltd., 57 Caledonian Road, London N1 9BU, UK, and
171 First Avenue, Atlantic Highlands, New Jersey 07716, USA
and:
The United Nations University, Toho Semei Building,
15-1 Shibuya 2-chome, Shibuya-ku, Tokyo 150, Japan
in cooperation with
The Third World Forum, B.P. 3501, Dakar, Senegal.

Translated by A. M. Berrett.

Cover designed by Andrew Corbett.
Typeset by Composer Typesetting, Bath, Avon.
Printed and bound in the United Kingdom
at The Bath Press, Avon.

British Library Cataloguing in Publication Data

Gakou, Mohammed Lamine
The crisis in African agriculture.—
(Studies in African political economy).
1. Food supply—Political aspects—Africa
I. Title II. United Nations. *University*
III. Series IV. La crise de l'agriculture
Africaine. *English*
338.1'9'6 HD9017.A2

ISBN 0-86232-732-6
ISBN 0-86232-733-4 Pbk

Library of Congress Cataloging-in-Publication Data

Gakou, Mohamed Lamine.
The crisis in African agriculture.

(United Nations University studies in African political economy)
Translation of: La crise de l'agriculture africaine.
Bibliography: p.
Includes index.
1. Agriculture—Economic aspects—Africa.
I. Title. II. Series
HD2117.G3513 1987 338.1'096 87-13355
ISBN 0-86232-732-6
ISBN 0-86232-733-4 (pbk.)

Contents

Tables

Acknowledgements

This book, published in the Studies in African Political Economy series under the general editorship of Samir Amin, is the outcome of research done by the author in the framework of the programmes of the Third World Forum (Intra-African Cooperation) of the United Nations University (African Regional Perspectives—Nation-building or Transnationalization in Africa?). We should like here to thank those institutions. We should add finally, as is usual, that the opinions expressed here are the responsibility of the author alone and do not represent the views of the organizations mentioned.

Foreword

In this book, we have endeavoured to demonstrate that the crisis in African agriculture affects virtually the whole continent, even if it is true that some countries are managing better than others. In the attempt to understand the causes of this crisis which is affecting not only agriculture but the whole society and economy, we have felt it necessary to go back to precolonial societies to see if their forms of organization and level of development could prevent the victory of the aggression mounted by the capitalist mode of production.

Capitalism having succeeded in ensuring its domination, we have attempted to grasp the forms of exploitation which developed during that era, and their consequences for the subject peoples. Facts and observation show clearly that there is no possibility of survival in the framework of the present world system. Is the alternative, which can materialize only in the form of a delinking from the system, possible today and on what conditions?

These are the issues we raised, without being sure of providing definitive conclusions. Societies need to be reorganized on the basis of transformations of class relations. Such a prospect is not foreseeable in the immediate future but it is the condition for the alternatives that can liberate the African peoples, make them masters of their fate and liberate their creative initiative.

M. L. Gakou

Independence did not alter this mode of integration into the world capitalist system. As Gakou shows (Chapter 4), independence came in response to the demands of the new stage of the globalization of capital (the rebuilding of Europe and the hegemony of the United States) and not in response to the African peasant problem. Moreover, the prosperity of the 1960s in the West led in Africa to a new euphoria for the 'extraverted system'—that system oriented to export markets. And while, courageously and clear-sightedly, René Dumont, ever sensitive to the peasant question, denounced the 'false start in Africa', the World Bank, which today weeps crocodile tears over the fate of the peasants (while its counterpart the International Monetary Fund makes the poorest foot the bill for the bankruptcy) sustained with overwhelming enthusiasm the policies that ten years later were to lead to catastrophe.

The crisis of the 1970s and 1980s results from the combined effects of the over-exploitation of the land and of men and women, raised to a level that can be raised no more, and the crisis engulfing the capitalist system as a whole. Faced with this the proposals that rain down on Africa at an ever-increasing rate are scarcely anything more than the expression of the 'quest for palliatives', as Gakou shows (Chapter 5).

If it is only a matter of palliatives, it is because the 'pro-agriculture' talk nurtured by the media in the West is contrasted with an alleged 'preference for industrialization' which is said to be the source of the bankruptcy. For the reason for seeking greater output per farmer is precisely in order to make possible a higher degree of urbanization. But urbanization without industrialization cannot but be parasitical and disastrous. Conversely, industry (but not just any industry) is necessary to make greater agricultural output possible: it has to supply it with machinery and offer it a growing market in return. That is what the option for an autocentred national and popular strategy means. If this option is rejected in favour of a systematic preference for integration into globalization (i.e., the world capitalist economy), talk of giving priority to agriculture becomes hollow and at worst demagogic. The World Bank Berg Report[1] has these characteristics, refraining as it does from making a self-criticism of the Bank which supported the policies that led to the present bankruptcy, refraining from making a critique of import-substitution industrialization (which finds its preferred market in the expansion of demand from the middle classes at the expense of the rural and urban popular classes) which it advocated, and refraining from analysing the implications of its proposal for export-oriented industry. But this derives its 'comparative advantage' from its low wages, which it thus contributes to reproducing. The contradictions of these 'proposals' are moreover obvious: export-oriented industry presupposes low wages and consequently, low food prices, at the very time that the raising of these latter to inspire the peasants to produce more is being recommended.

The populist dress that MacNamara, when he was President of the World Bank, gave these proposals does not alter their implications. The themes of 'basic needs' and the strategy of 'family smallholdings' analysed by Gakou scarcely conceal the worst choice scenario which unfortunately some confused sections of the Western left have confused with the popular interest. In any case, these oratorical flourishes have never prevented Western 'aid' agencies from giving *de facto* preference to support for agri-business and the kulaks—in the name of efficiency. The fact that these policies continue to be advocated in the West testifies basically to the lack of seriousness with which Africa is treated. In fact, Africa, in the imperialist vision of the world, is above all for the West a source of mineral resources. Neither its industrialization nor its agricultural development are thus really taken seriously.[2]

Nothing in nature ordains that African agriculture should be impoverished. Doubtless the under-population of tropical Africa (compared with the dense population of tropical Asia) constitutes an obstacle to intensification which calls for major internal population movements. And whatever has been said, the Sahel itself is not 'poor'. Here there is water (a series of rivers whose flow is equal to that of the Nile, exceptional underground water layers, according to studies that have been kept secret[3]), sources of energy (is not uranium one of these sources? And the sun? And oil less than a thousand metres down?), land suitable for cultivation, and people. A social system that proves itself incapable of co-ordinating these 'factors' in a satisfactory plan able to feed the people involved does not deserve to be described as rational. Let us recognize then that the capitalist system is not rational since it does not necessarily guarantee the reproduction of the labour force in each of its segments. Here, in the Sahel, for capitalism such as it is, it is the existence of the Sahelian peoples that is 'irrational'. For this capitalism, things would be more profitable if there was only uranium in the Sahel and no useless Sahelians. Such is the logic of the world system for which Africa is still exclusively a source of raw materials. From their endless stress on emergency 'relief' distributions, Western and European institutions have created the impression that the Sahel was irrevocably condemned. Thus we see a certain progressive institution accept as self-evident, in a study of 'energy prospects' for the region, that since uranium is not intended for the 'natives', the Sahelians must be taught better ways of gathering brushwood in the desert and how not to waste it in Malagasy stoves! So let Africa adapt itself to the wastefulness of the West. Is there a better expression of the fate as mineral supplier to which imperialism consigns the continent and of the subordination of all so-called development programmes to this essential logic than this naive acceptance of the 'imperatives' of exporting the energy resources of the region? But why not the opposite; why should not Africa recover control and use of its resources, and Europe adjust itself to that?

The capacity of capitalism in the abstract to 'solve the problem of African development' could be discussed *ad infinitum*. Concrete capitalism, such as it actually exists, that is, globalized, not only has not 'solved' this problem (it even created it) over the last 150 years (or even over the last 400 years since the slave trade), but envisages nothing for the next 50 years. The challenge will therefore only be taken up by the African peoples, the day that the necessary popular alliances enable them to delink their development from the demands of transnationalization.

Samir Amin

Notes

1. For a critique of this World Bank report, see Samir Amin, 'Une stratégie de développement autocentrée est-elle possible pour l'Afrique?'; Kwame Amoa, 'Some problems of autocentered development in Africa'.
2. Fayçal Yachir, *The Struggle over Africa's Minerals: What is at stake?*, forthcoming in this series.
3. Explanatory notes and planning maps for the exploitation of the underground waters in the Sahel, Bureau de recherche géologique et minimânt, 1975.

Introduction

Our aim in undertaking this work is to demonstrate, or provide further confirmation that the crisis affecting Africa particularly — even though it is more widespread — has its profound roots in the integration of African economies into the world capitalist system. The agricultural sectors and the rural areas are most often the ones most affected because of this integration. The case of agriculture, which, in most countries, is in crisis because it is essentially oriented towards the world market and not towards the feeding of the local people, shows that it is idle for the underdeveloped countries, and particularly for Africa, to seek solutions to their problems in the framework of a system whose *modus operandi* and rules of the game operate in such a way that it is always the poorest and economically weakest that suffer the most serious consequences of the crisis. If the developed capitalist countries can make the underdeveloped countries bear at least a part of the burden of their own crisis, in these countries and in Africa in particular, the so-called 'non-modern', 'traditional' sectors, agriculture above all, bear more of the burden. Other explanations can be found for the crisis, but we feel that these explanations can be no more than secondary, the fundamental cause being the integration of Africa into a system over which it has absolutely no control.

Even in the Sahelian region there are reports of granaries of cereals always full during the precolonial period despite the low level of development of productive forces. But was it not this low level of development of productive forces that ultimately made Africa the victim of the capitalist mode of production? A brief look at the work of distinguished researchers who have studied precolonial African societies suggests that these societies were not adequately prepared to defeat the aggressions of capitalism despite great capacities for resistance often linked to very advanced levels of political and social organization. The long era of domination that followed saw Africa drained of its human and material substance which was sucked out by the invaders.

We have not sought in this study to present a specialist piece of work, but to say things simply and to recall truths which may appear obvious but which are increasingly glossed over and rejected because of their very simplicity. Yet these truths still remain highly relevant and fundamental.

Despite the still appreciable potential of African agriculture, hunger, malnutrition and poverty have got worse over the years, reaching the point of explosion in the 1970s. Absurd policies, breath-taking in their lack of imagination, continue to be implemented almost everywhere in Africa under the control of the system and its agencies (World Bank, IMF) by leaders who clearly stand to gain by them, but who crush the peoples. What alternative can one think of in the face of this ever-worsening situation? In our opinion, the solution can first of all only be political. All African and extra-African energies must endeavour to make the African peoples the true masters of their destiny through reorganizations of society that will ensure them all their rights and guarantee them all their freedoms. These are the prerequisites, and success requires constant and militant struggles.

Preface

Mohamed Lamine Gakou here gives an overview of the 'agricultural question' in contemporary Africa. This is a difficult task if the writer wishes both to respect the variety of situations on this vast continent and avoid the detailed juxtaposition of case studies, both to spare the reader the jargon of specialists and remain brief without lapsing into superficiality. But it is a task in which the author has succeeded.

Africa's agricultural failure—this continent is the only one in which per capita agricultural production is falling—is certainly widely recognized, but the explanations for it remain generally partial and contradictory. Does the distant past (precolonial Africa (Chapter 2)) bear some share of responsibility? If there is a 'specificity'—over and beyond their great variety—in the modes of organization of the rural areas of most of Africa, it is probably that the still virtually untouched communal or tribute-paying forms involved an extensive occupation of the land. This made possible a much higher level of food self-sufficiency than is often supposed thanks to a relatively high labour productivity (which went hand in hand with very low yields per hectare). A higher per capita production demands the shift to intensive modes calling for a considerably higher overall quantity of annual labour. This increase in per capita production is thus accompanied by a lower productivity of labour (of physical production per 'day' of labour) but also by a higher yield per hectare. The shift to intensive agriculture, the *sine qua non* of any development worthy of the name, constitutes the challenge that the African peoples will pick up.

But the challenge has not yet been picked up. Colonization not only failed to pick it up; it never set out to do so. It was easier for it, as Gakou shows (Chapter 3), to secure an immediate super-profit at no cost (involving no investment) by forcing the peasants of Africa to perform unpaid—or very poorly paid—surplus labour through forms of indirect control (Chapter 6). A slightly higher per capita production at the price of rather more work, without machines, or modern inputs (but destroying the soil of Africa), combined with a deterioration in the conditions of peasant life were enough to elicit a considerable profit for the capital dominating the global system. Colonization thus continued the old tradition of the slave trade; exploitation by pillage that guaranteed neither the long-term reproduction of the labour force, nor the reproduction of the natural conditions of production.

remained distinctly positive. Despite significant declines compared to the previous period, the Eastern countries again recorded the highest growth rates in the group of developed countries. During this decade, the underdeveloped market economy countries still recorded total production growth rates higher than those of the developed market economy countries. But their per capita production was insignificant not only compared to those of the developed countries, but also compared to their own previous performances because of general declines in total production (except for that of Africa, which recorded a significant recovery), and because of the increase in population growth rates compared to the previous period.

Table 1
Annual Growth Rate of Food Production, Total and Per Inhabitant

	1952-62		*1961-70*		*1970-78*	
	Total	*Per inhab.*	*Total*	*Per inhab.*	*Total*	*Per inhab.*
World	3.1	1.1	2.9	0.9	2.5	0.8
Developed market economies	2.5	1.3	2.6	1.5	2.2	1.3
Western Europe	2.9	2.1	2.4	1.5	1.6	1.2
North America	1.9	0.1	2.3	1.1	2.9	2.0
Eastern Europe +USSR	4.5	3.0	3.8	2.7	2.6	1.8
Developing market economies	3.1	0.7	2.8	0.2	2.7	0.4
Africa	2.2	0.0	2.9	0.4	1.4	−1.3
Latin America	3.2	0.4	3.1	0.2	3.0	0.2
Asia	3.1	0.8	2.6	0.1	2.7	0.1
Asian centrally planned economies	3.2	1.4	2.9	1.0	2.8	1.1

Sources: 1952-62, UN; 1961-70 and 1970-78, UNCTAD.

With regard to Africa, which had experienced a recovery in production whereas other regions had suffered declines, the trend averaged around a very low annual growth rate in per capita food production of 0.2% over the period. The planned economy countries of Asia experienced higher rates of 0.7% because their population growth rates remained distinctly lower than those of other underdeveloped regions.

For the third decade, it can be observed that the trends initiated during the second decade were accentuated. For the developed countries, whereas in North America the annual growth rate in total food production rose, in Western Europe and the Eastern countries rates continued to decline.

Overall, the growth in total food production was even stronger in the underdeveloped market economy countries than in the developed ones. But as the demand for food had continued to rise in the former, the situation remained generally stagnant in comparison to the previous decade.

Only Africa, which during the second decade had been able to achieve growth rates roughly equal to those of other underdeveloped market economy regions, experienced a sharp decline, its per capita food production growth rate falling to −1.3%. In the underdeveloped world as a whole, the socialist countries of Asia continued to register the best performances in per capita food production.

Obviously, many factors have to be taken into account in explaining the evolution of these trends: historical, socio-political and economic factors, conjunctural problems, unforeseeable disturbances, irregular occurrences, etc. In these conditions, we can do no more here than make a few observations on the particular circumstances that may have influenced this or that region, more detailed explanations of the evolution noted requiring a much more ambitious project than the present work.

The decade 1950-60 corresponds to the immediate post-Second World War period. Since this war was fought worldwide, the economic structures of most regions of the world were seriously affected, even though unequally. The countries that were most seriously affected, but had powerful potential means to face up to the task of post-war recovery, were the developed countries. Among these countries, the European ones suffered most destruction. Emergency measures had to be taken for economic reconstruction, beginning with facing up to famine and food shortages. Solving food supply problems was made easier for them because of the existence of an industrial infrastructure which made it possible to supply agriculture with the taxes it needed. For those countries it was simply a matter of adopting a policy to promote the development of agricultural production.

In the Soviet Union, where agriculture was very backward (by 1950 it had not yet succeeded in recovering its pre-war level), vigorous measures were taken. Their aim was to loosen the constraints affecting the private sector so as to increase production substantially. Income taxes on private agricultural incomes were cut by a half. Compulsory deliveries to the state were suspended. Since industry was unable, in the case of the USSR, to respond at once to the tax expectations of agriculture, the authorities decided to develop agriculture extensively through the opening up of new lands. 'In all, the bringing into cultivation of the virgin lands affected 36 million ha. in 1954-56 of which 20 million were in Kazakhstan—representing an increase of 23% in the cultivated area in two years.'[1]

Numerous measures were adopted to provide economic incentives. These included sharp rises in the prices of agricultural products, and especially of foodstuffs; tractor and machinery stations were managed by members of kolkhozes; the kolkhozes were grouped together to reduce inequalities in the countryside.

While the private and co-operative sectors were strongly encouraged, the state sector was not neglected. The area covered by state farms rose from 15 million ha. in 1953 to 50 million ha. in 1957. These series of measures enabled marketed production of cereals to rise by 75% in five years, with increases of 88% for milk and 50% for meat.

The other countries of Europe which lacked the arable land potential of the USSR stepped up production by increasing the products of industries upstream from agriculture, and by adopting vastly improved techniques tested notably in Denmark and Holland. Thus, in France, beef production doubled between 1948 and 1967 and milk output rose by 80%.

It is, therefore, not difficult to understand that agricultural growth rates in the countries of Eastern and Western Europe were higher than those in North America during the first post-war decade because the former had been much more seriously affected, and the need for the recovery of growth was more pressingly felt.

As for the underdeveloped countries, the particularly remarkable slowness of the growth of African food production compared to other regions is doubtless to be explained—as we shall see below—by the fact that the period 1950-60 corresponded to the flourishing period of the colonial economy which was only interested in export agriculture, promotion of which was to assist the recovery of the metropolitan economies. It should, however, be added that statistical services were practically non-existent in Africa at that time and that the data on food production in particular are very unreliable.

The reversal of the trend that appeared in the developed countries during the second decade, and which saw a sharp rise in growth rates in North America while they fell in Europe, might be linked first to the scale of the increase in world demand for food products and then to a greater capacity of American agriculture to adapt to internal and external market conditions because of its vast potential. Furthermore, significant government subsidies were forthcoming under pressure from the powerful big farmers' lobby, the American Farm Bureau Federation. These subsidies went mainly to the wealthy farmers, the owners of the largest and most modern holdings. President Eisenhower said on this subject in 1960: 'The main beneficiaries of our price support policies have been the two million largest, most highly mechanized production units'.[2]

As a result of the absence of class solidarity between large and small farmers, therefore, pressure groups representing the interests of rich farmers—which were the only ones to get the attention of the political classes—succeeded in securing that all government expenditure on agriculture go to the rich farmers, thus condemning the poorest to disappear.[3] Thus, the largest, most mechanized holdings, that were also the sole beneficiaries of government assistance, were fully able to face up to the conditions of the international market.

Unlike America, Europe had a feudal past with intense social struggles in the countryside which led to the strengthening of class feeling. Capitalist society which followed feudal society saw the rise of a powerful and well structured labour movement. In order to prevent a class alliance between this movement and the mass of small peasants against the whole bourgeois system, the European big rural bourgeoisie, as well as the established authorities, had an interest in seeking the support of this mass. They could not therefore take the risk of embarking on a process of brutally eliminating smallholders. These socio-historical conditions meant and still mean that the agrarian structures in Europe differ from those in America. Thus, despite the existence of an extremely large world market for agricultural products, and despite the extraordinary development of agri-business to which large holdings ought to adapt better, Western European agriculture still remains heavily marked by the past and rests principally on smallholdings.

As for the USSR, the stress put on the development of extensive agriculture bore fruit between 1950 and 1960. But after that, the phenomenon of decreasing yields was strongly felt and growth rates began to decline.

During the 1970s, the broad trends of the previous decade became more marked; North America, because of its enormous potential and its agrarian structures, increased its production and its domination of the world market for food products over Western Europe, the USSR and the European socialist states.

During the 1960s and 1970s, the underdeveloped regions (except for Africa in the 1970s) maintained total production growth rates relatively stable around 3% per annum, exceeding those of the developed countries as a whole.

The complexity of the situation in Latin America makes any attempt to link these performances to precise causes hazardous, numerous factors coming into play depending on the country: climatic conditions, development of commercial production in extensive agriculture in large latifundia estates, implementation of agrarian reforms in some countries, etc. On the other hand, in Asia where population density is very high, pressure on the land is such that only the increase in yields through more intensive cultivation and the use of selected varieties have made it possible to achieve these performances. In Africa the sharp decline in food production during the 1970s compared to the 1960s is, first of all, the consequence of the conditions of exploitation of previous decades. Of course, there were in addition the aggravating circumstances of the very considerable climatic variations during the 1970s, while the rising costs of inputs and the inadequacy of irrigated areas prevented any possibility of intensifying production.

But however high growth rates in food production in the underdeveloped regions were compared to those in the developed countries, the former's per capita food production still remained much lower because of high population growth. Since socio-cultural structures in the underdeveloped countries give no grounds for hoping for a rapid reduction in population growth rates, the solution can only be sought in raising total food production even higher even though that will not guarantee an equitable redistribution of food intakes for each stratum of the population.

These few observations are far from constituting an attempt to explain the evolution of the trends over three decades in the various regions of the world.

In fact, such an attempt at explanation should proceed essentially from the study of class relations and their evolution in various societies under the impact of the domination of the world capitalist system.

The work of attempting to understand the evolution of agriculture through the forms of the domination and exploitation of the producers will be attempted here for Africa only, since that is the subject of our study.

But before continuing, we should complete this section which puts food production in Africa in the context of other regions through an attempt to compare potentials and levels of development, at least for the underdeveloped regions of the world.

Potentials and levels of development

Compared to other regions of the world, both developed and underdeveloped, Africa has considerable potential arable land resources, as Table 2 shows. Indeed, potential arable land is estimated at 733 million ha. as against 570 million ha. in Latin America and 628 million ha. in Asia.

With 10% of the world's population, Africa has 25% of world potential arable land, that is 'land that appears to have economic potential for arable agriculture under existing levels of technology.'[4] And only 27.8% of arable land is under cultivation. But techniques of production, which are rudimentary and rest, in most cases, on shifting agriculture, involve the significant waste of land resources and the accelerated destruction of soils. Thus, because of its low technical levels, African agriculture lags considerably behind that of other underdeveloped regions.

The continent suffers a serious water problem and rainfall is frequently irregular over much of the region. Irrigation, which could have provided solutions to this tragic problem, is unfortunately very little developed. In the continent as a whole, only two countries (Egypt and Sudan) have major irrigated areas. Table 3 shows clearly how far Africa lags behind other underdeveloped regions. In millions of hectares, irrigated areas in 1962 totalled only 1.1 south of the Sahara as against 44.1 in Asia and the Far East, 10.6 in Latin America and 16.7 in the Middle East. According to projections for 1985, irrigated areas would not exceed 1.9 million ha. south of the Sahara despite considerable surface and underground water resources. The same lag can be observed in the use of inputs making it possible to intensify production. Thus, Africa produced only 0.3 million tons of fertilizers in 1967-68, and 0.9 in 1972-73. As for consumption, it rose from 0.4 million tons in 1967-68 to 0.9 in 1972-73. Table 4 brings out the gap compared to other underdeveloped regions. In terms of arable area, consumption of fertilizers in the region was only 5.7 kg./ha. as against 11.3 in Latin America and 18.7 in Asia in 1967-68. In 1972-73, the figures were respectively 10 kg., 27.4 kg. and 28.6 kg. By way of illustration, in the United States the consumption of fertilizers is more than 85 kg./ha.

If we measure the level of mechanization by the quantity of energy consumed from all sources, it is estimated that Africa has available 0.05 hp./ha. of arable land as against 0.19 in Asia and 0.27 in Latin America.

1: The Performance of African Agriculture, 1950–1980

Basic data and broad trends

Agricultural production and food supplies

Overall, during the three decades from 1950, total agricultural production as well as total food production rose substantially in all regions of the world, both developed and underdeveloped. According to OECD statistics, total food production in the developed market economies rose by 115% between 1950 and 1975, while the developing market economies achieved a better performance with an increase of 130% over the same period.

These remarkable increases did not, however, have the same effects on per capita food supplies in the developed regions and in the underdeveloped ones. Whereas in the former, per capita food supplies were satisfactory given low population growth rates, in the latter, population increases were quite often close to production increases making rates of growth of per capita food production virtually insignificant.

Table 1, which shows growth rates for the various regions, brings the trends out. The pattern has varied from region to region and decade to decade.

Thus, during the first decade, 1952-62, except for Africa which only achieved a growth rate of 2.2%, all the other regions of the underdeveloped world achieved growth rates in total food production of above 3%, rates higher than those of the developed market economies which averaged 2.5%. But rapid population growth in the underdeveloped countries reduced per capita food production to levels below those of the developed countries. During that decade in the developed countries as a whole, the Eastern countries recorded the highest rates, both for total production and for per capita production. North America had the lowest rates. In the developing countries as a whole, the planned economies of Asia recorded the highest growth rates while Africa recorded the lowest, far lower than those of other underdeveloped regions.

The second decade, 1962-72, was, in general, marked by a slowing down of the growth of production in both the developed and the underdeveloped world. For total food production, except for North America which recorded a striking recovery, all the other developed regions recorded decreases, sometimes quite significant ones. Nevertheless, their per capita food production

Table 3
Proposed Irrigation Development

	Irrigated Area						Growth Rates in Irrigated Area	
	Arable Area			Harvested Area			Arable	Harvested
	1962	1975	1985	1962	1975	1985	1962-1985	
	(Million Ha.)			(Million Ha.)			(% p.a.)	
Africa South of the Sahara	1.1	1.5	1.9	1.1	1.6	2.1	2.4	2.9
Asia and Far East	44.1	55.5	68.1	49.5	75.0	102.7	1.9	3.2
Latin America	10.6	13.5	17.4	18.2	11.4	16.2	2.2	3.0
Near East and N.W. Africa	16.7	18.5	20.1	12.8	16.1	17.8	0.8	1.5
Total	*72.5*	*89.0*	*107.5*	*81.6*	*104.1*	*138.8*	*1.7*	*2.9*

Source: *Provisional Indicative World Plan for Agriculture.*

In Western Europe, the quantity of energy per ha. is 0.93.

In conclusion, it can be said that compared to other regions of the world, Africa has available major natural resources for the development of agriculture. But these resources are both under-exploited and poorly exploited because of the very low level of development of the region. The result is very low yields and productivities. Calculated from the FAO's statistical yearbooks, average yields in Africa, as a percentage of world averages, were the following for the main crops in 1948-52: wheat 72%, rice 60%, millet and sorghum 105%, maize 55%, groundnuts 82%, cotton 42%. In 1974 the yields had declined steeply compared to world averages: wheat 45%, rice 55%, millet and sorghum 95%, maize 44%, groundnuts 71%. Only cotton yields increased, from 42% to 44%.

What is more, this low level of yields was not compensated for by a proportionate increase in cultivated areas despite the existing potential, at least for crops intended for consumption. Thus, between 1948-52 and 1972-74 the area sown to the main cereal crops in Africa rose by only 15% whereas for the same period and the same crops, it rose by 24% in the world as a whole, 25% in the Far East and 64% in Latin America.

On the other hand, if we look at cash crops, the extension of cultivated area in Africa was, in percentage terms, considerably higher than in the rest of the world. In cotton production, area increased by 70% in Africa as against 9% in the Far East. In groundnut production, it rose by 65% in Africa as against 63% in the world as a whole and 55% in the Far East.

Consequently, the share of Africa in world production of food crops has tended to fall whereas its share in the production of export crops has risen substantially (see Table 5).

Table 4
Production and Consumption of Chemical Fertilizers,[1] World and Main Regions (July-June 1967-68, 1971-72, 1972-73)

	Production			*Growth rate*	*Consumption*			*Growth rate*
	1967-68	1971-72	1972-73	1967-68	1967-68	1971-72	1972-73	1967-68
	(millions of tons)			(% p.a.)	(millions of tons)			(% p.a.)
Developed market economies	40.4	46.5	48.5	3.7	33.8	38.7	40.5	3.7
North America	17.7	21.3	21.9	4.3	14.5	16.5	17.2	3.5
Western Europe	17.9	19.9	20.8	3.0	15.1	18.1	18.8	4.6
Oceania	1.3	1.3	1.4	2.0	1.5	1.4	1.6	2.2
Eastern Europe and USSR	14.1	21.7	23.3	10.5	12.6	18.5	20.1	9.8
Total industrialized countries	*54.5*	*68.2*	*71.8*	*5.7*	*46.4*	*57.2*	*60.6*	*5.5*
Developing market economies	2.5	5.2	6.1	19.6	5.8	9.9	11.5	14.7
Africa	0.3	0.8	0.9	24.9	0.4	0.8	0.9	17.6
Far East	1.1	2.3	2.6	18.8	2.3	4.7	0.3	18.3
Latin America	0.8	1.3	1.4	11.9	2.0	3.1	3.8	13.7
Near East	0.4	0.8	1.1	22.5	0.8	1.3	1.5	13.4
Asian planned economies	1.9	3.5	4.0	16.1	2.7	4.9	5.3	14.5
Total developing countries	*4.6*	*8.7*	*10.1*	*17.0*	*8.5*	*14.8*	*16.8*	*14.6*
World Total	*59.1*	*76.9*	*81.9*	*6.8*	*54.9*	*72.0*	*77.4*	*7.1*

[1] Content in nitrogen, phosphorus and potash.

Source: OECD.

Agricultural situation

While the situation of African agriculture is quite critical overall, there is nevertheless a certain disparity of situations between countries or groups of countries. Thus, during the decade 1960-70, of 50 countries, 17 increased their per capita food production, with increases ranging from about 5% to 50%. The most striking performances were achieved by Swaziland (over 50%

increase), Tanzania (over 40%), Burundi, Libya, Malawi, Sudan, Ivory Coast, Rwanda, Zaire and Cameroon. Over the same decade 20 countries experienced significant declines in production varying from 5% to 35%. The most significant declines were recorded in Senegal (a decline of almost 35%), Chad 25%, Congo 20%, Mali, Nigeria, Algeria (about 15%). Thirteen countries remained relatively unchanged.

Table 5
Africa's Share in World Production of Food and Export Crops (in %)

	1948-52	*1971-72*
Maize	7	10
Millet and sorghum	17	13
Wheat	2.7	2.6
Rice	2.1	2.5
Barley	6	3
Potatoes	0.5	0.8
Sweet potatoes and yams	23	16
Cassava	50	40
Coffee	12.5	27
Cocoa	65	72
Tea	3	9
Tobacco	5	6
Cotton	9	11

Sources: Calculated by the author from data in FAO production yearbooks.

During the decade 1970-80, the situation deteriorated sharply. Of 47 countries for which statistics were available, only seven had increases in per capita production ranging from 5% to 30% approximately: Tunisia, Libya, Cameroon, Zambia, Sudan, Burundi, Ivory Coast. Twenty-nine countries had declines of between 5% and 25%.

The worst performances were recorded in the following countries: Mauritania, Ghana, Togo, Morocco, Nigeria, Algeria, Ethiopia, Niger, Angola, Mali, Uganda (see FAO yearbooks).

Table 6 summarizes the evolution of the main sectors of the economy by country for the periods 1960-70 and 1970-76. It shows the annual average growth rates by sector as well as the contribution to GDP. This table confirms the deterioration of the situation during the 1970s compared to the 1960s. Of the 37 countries listed in the table about 20 saw their average annual agricultural production growth rate fall from one period to the other which resulted to some extent in a smaller contribution by the agricultural sector to GDP, whereas the considerable proportion of the population working in the sector remained almost unchanged. This fall led in most countries to a significant gap between production and consumption of agricultural products, as Table 7 shows for cereals. It can be observed that for all the countries mentioned, between 1969-71 and 1978, self sufficiency in cereals was generally

Table 6
Evolution of the Main Sectors of the Economy by Country (1960-70, 1970-76)

Average annual growth rate—Contribution to GDP by sector

	Agriculture				*Industry*				*Services*			
	Growth rate		*Cont. GDP*		*Growth rate*		*Cont. GDP*		*Growth rate*		*Cont. GDP*	
	1960-70	70-76	60-70	70-76	60-70	70-76	60-70	70-76	60-70	70-76	60-70	70-76
1. Ethiopia	2.2	0.9	65	50	7.4	1.6	12	15	7.3	4.4	23	35
2. Mali	1.3	−0.8	55	38	4.0	8.9	10	17	4.4	5.5	35	45
3. Rwanda	—	3.3	81	52	—	8.4	7	22	—	3.5	12	26
4. Somalia	−1.5	−1.2	45	31	3.3	10.3	17	8	2.1	8.0	38	61
5. Burkina Faso	0.0	3.2	55	34	3.8	7.0	13	19	2.3	1.8	32	67
6. Burundi	—	1.0	—	64	—	4.3	—	15	—	1.1	—	21
7. Chad	1.8	−1.3	55	52	3.9	8.1	12	14	2.9	−0.6	33	34
8. Benin	—	−0.3	—	39	—	9.8	—	20	—	6.0	—	41
9. Malawi	2.9	5.5	58	45	13.9	12.4	11	22	8.9	11.4	31	33
10. Zaire	3.9	1.9	30	16	35.9	5.0	27	30	−2.5	5.0	43	54
11. Guinea	2.1	10.2	—	43	6.2	3.9	—	33	2.2	3.2	—	24
12. Niger	3.3	−4.0	66	47	11.1	10.0	10	24	0.6	0.8	24	29
13. Lesotho	—	—	73	38	—	—	—	8	—	—	—	54
14. Mozambique	2.1	2.1	55	45	10.8	−3.8	9	15	5.8	−2.1	36	40
15. Tanzania	3.7	2.5	57	45	8.0	2.9	11	16	5.3	2.8	32	32
16. Madagascar	—	1.2	37	29	—	2.0	10	20	—	4.5	53	51
17. Sierra Leone	1.4	2.0	—	32	2.7	−30	—	23	4.2	4.0	—	45
18. Central Afr.Rep.	0.8	1.9	45	37	5.5	4.7	12	23	0.1	−1.8	43	40

Table 6 (continued)

	Agriculture				*Industry*				*Services*			
	Growth rate		*Cont. GDP*		*Growth rate*		*Cont. GDP*		*Growth rate*		*Cont. GDP*	
	1960-70	70-76	60-70	70-76	60-70	70-76	60-70	70-76	60-70	70-76	60-70	70-76
19. Kenya	5.9	1.6	38	30	7.5	9.8	18	23	7.9	5.1	44	47
20. Uganda	2.8	1.3	52	55	7.8	−6.7	13	8	8.3	−3.2	35	37
21. Togo	4.3	3.0	55	25	7.3	7.0	16	21	8.8	3.7	29	54
22. Egypt	2.9	3.0	30	29	5.4	4.3	24	30	6.1	13.4	46	41
23. Cameroon	6.5	3.4	48	33	7.7	3.3	10	20	11.1	0.7	42	47
24. Sudan	3.3	8.8	58	41	1.7	2.8	15	16	−2.2	7.5	27	43
25. Angola	4.0	−0.7	50	29	9.8	11.6	8	27	3.9	3.0	43	44
26. Mauritania	2.4	−2.1	57	35	15.8	7.1	21	37	13.0	−1.0	22	28
27. Nigeria	−0.5	−0.2	63	23	13.8	12.6	11	50	5.2	9.5	26	27
28. Senegal	1.9	3.4	30	28	3.7	3.9	20	24	2.5	−0.1	50	48
29. Zambia	2.0	3.2	11	14	−0.1	3.4	63	41	8.1	4.4	26	45
30. Liberia	6.3	4.9	40	29	7.8	0.3	37	37	2.6	9.4	23	34
31. Congo	4.6	−7.2	16	15	7.6	22.6	18	43	2.4	7.0	68	42
32. Morocco	4.2	0.6	29	21	4.2	7.8	24	31	3.9	5.5	47	48
33. Zimbabwe	—	—	18	16	—	—	35	40	—	—	47	44
34. Ghana	3.7	1.3	41	49					−1.4	3.8	40	26
35. Ivory Coast	4.2	3.5	43	25	11.6	7.9	14	20	10.0	7.7	43	55
36. Tunisia	2.0	9.2	24	21	8.7	10.1	18	30	2.9	9.7	58	49
37. Algeria	−1.6	−8.7	21	7	10.5	16.4	24	57	2.3	−4.6	55	36

Table 2
World Arable Land Resources (in Millions of Hectares)

			Asia			*North America*			*Europe*		
	Latin America	*African Total*	*Total* (excl. USSR incl. China)	*China*	*Oceania*	*Total*	*USA*	*USSR*	*Total*	*East*	*West*
Arable land under cultivation (base year)	119(a)	214(a)	467(a)	130(a)	47(b)	230(c)	191(d)	225(e)	141(a)	42(f)	99(a)
Projected arable land—1985	148	242	484	143	71	224	185	230	137	41(f)	96
Potential arable land(g)	570	733	628	200	107	495	456	270	141	45(f)	96

a) Base = 1970
b) Base = 1970-71
c) Base = 1971
d) Base = 1969
e) Base = 1973
f) Excluding USSR
g) *Potential* arable land is broadly defined in this sense as land that appears to have economic potential for arable agriculture under existing levels of technology.

Note: The definition of what exactly constitutes arable land varies somewhat between these estimates according to the severity of climatic, terrain and soil constraints that have been imposed. For this reason it is not possible to make precise comparisons between regions or to give a world total.

Source: OECD, *Studies of Trends in World Supply and Demand of Major Agricultural Commodities.*

Table 7
Africa: Visible Per Capita[1] Production and Consumption of Cereals in Certain Countries

	Production				*Visible consumption*			
Sub-region and country	1969–71	1972–74	1975–77	1978	1969–71	1972–74	1975–77	1978
	Average Figure (kg./p.a.)				Average Figure (kg./p.a.)			
North Africa								
Algeria	141	118	101	146	177	254	224	281
Morocco	300	266	226	246	321	312	298	344
Total	*187*	*172*	*161*	*184*	*215*	*231*	*226*	*266*
Sahel								
Mali	197	138	191	220	205	165	204	227
Niger	317	181	261	298	305	190	262	311
Burkina Faso	187	171	193	186	191	181	198	206
Total	*196*	*143*	*174*	*194*	*214*	*174*	*198*	*233*
West Africa								
Guinea	175	141	142	140	185	156	157	159
Togo	151	109	114	132	159	119	122	147
Total	*118*	*102*	*106*	*106*	*127*	*113*	*120*	*131*
Central Africa								
Angola	101	85	79	79	90	93	98	92
Central Afr. Rep.	58	64	49	42	65	73	55	45
Total	*42*	*38*	*38*	*34*	*48*	*54*	*54*	*49*
East and Southern Africa								
Lesotho	199	170	154	172	234	229	202	271
Tanzania	104	82	92	97	107	97	106	108
Total	*151*	*154*	*145*	*141*	*158*	*162*	*154*	*150*
Other countries[2]								
Egypt	196	190	186	186	230	272	277	313
Libya	56	89	104	113	239	252	277	325
Total	*181*	*178*	*173*	*173*	*222*	*260*	*268*	*298*

[1] Food and non-food uses.
[2] Invited members of the UN Economic Commission for Africa but not belonging to the regional conference.

Source: United Nations.

not achieved. Thus net imports of cereals for the countries of the region were considerable, as Table 8 shows with data for 1969-71 to 1978.

For the period under review imports of cereals more than tripled in North Africa. Those of Morocco and Algeria quadrupled. They quadrupled in the Sahel zone. Those of Chad increased almost ninefold. Niger, which exported almost 49,000 tons in 1969-71, imported 64,000 tons in 1978. For the rest of

Table 8
Africa: Net Cereal Imports (000 tons)

Sub-regions and countries	1969-71	*ACTUAL* 1972-74 Average Figure	1975-77	1978	*Alternative Projections* 1985	
North Africa						
Algeria	482	1,986	1,975	2,307	2,280	2,168
Morocco	308	770	1,302	1,872	2,395	1,693
Sudan	167	95	66	102	–168	–308
Tunisia	442	12	474	876	756	492
Total North Africa	*1,399*	*3,163*	*3,817*	*5,157*	*5,263*	*4,045*
Sahel						
Cape Verde I.	36	39	35	69	39	34
Chad	10	28	16	87	322	322
Gambia	15	16	36	100	34	34
Mali	42	149	76	41	257	204
Mauritania	64	126	121	150	186	176
Niger	– 49	39	5	64	24	– 45
Senegal	291	357	443	501	636	507
Burkina Faso	24	54	30	122	299	306
Total Sahel	*433*	*808*	*662*	*1,134*	*1,797*	*1,538*
West Africa						
Benin	20	29	39	81	87	55
Cameroon	86	113	83	119	181	211
Ghana	114	168	158	333	108	193
Guinea	41	61	55	94	134	8
Guinea-Bissau	28	30	28	30	29	21
Ivory Coast	141	213	162	370	320	343
Liberia	52	54	51	67	85	87
Nigeria	335	463	912	1,887	2,511	2,356
Sierra Leone	62	67	36	73	108	94
Togo	16	21	19	36	63	49
Total West Africa	*895*	*1,219*	*1,543*	*3,090*	*3,626*	*3,417*
Central Africa						
Angola	– 65	47	127	154	89	152
Central Afr. Rep.	12	16	10	5	29	43
Congo	29	35	32	67	73	78
Gabon	6	19	47	29	38	42
Sao Tome and Principe	6	7	6	9	8	8
Zaire	183	402	356	314	950	910
Total Central Africa	*171*	*516*	*578*	*578*	*1,187*	*1,233*

Table 8 (continued)

Sub-regions and countries	1969-71	1972-74	1975-77	1978	Alternative Projections 1985	
	ACTUAL Average Figure					
Eastern and Southern Africa						
Botswana	54	70	41	53	100	69
Burundi	10	16	9	15	64	45
Comoros	16	15	13	18	23	22
Ethiopia	51	21	56	122	1,227	987
Kenya	−94	83	30	99	403	227
Lesotho	36	66	56	121	94	65
Madagascar	18	100	117	260	331	301
Malawi	32	25	27	—	27	89
Mauritius	125	141	142	245	166	174
Mozambique	93	60	193	68	288	258
Swaziland	—	—	38	40	—	—
Tanzania	8	10	11	10	52	20
Uganda	6	7	7	10	11	11
Zambia	22	226	232	184	352	264
Total Eastern and Southern Africa	*617*	*776*	*1,062*	*1,073*	*3,702*	*2,899*
Total Africa	*3,515*	*6,482*	*7,662*	*11,032*	*15,575*	*13,132*
Other countries[1]						
Djibouti	16	23	27	20	35	36
Egypt	1,123	2,938	3,504	5,111	5,278	4,805
Libya	367	370	550	570	575	433
Somalia	75	61	115	85	153	91
Total other countries	*1,581*	*3,392*	*4,196*	*5,786*	*6,041*	*5,365*

[1] Invited members of the UN Economic Commission for Africa but not belonging to the regional Conference.

West Africa excluding the Sahel countries, imports tripled. Those of Nigeria increased fivefold, those of Benin fourfold. They also tripled in Central Africa. Angola, which used to export 10% of its production, came to import 21% of its consumption, and Gabonese imports went up fivefold.

Overall, it was eastern and southern Africa that had the lowest imports, the figures having not even doubled. There too, however, some countries faced particularly serious situations. Indeed, Tanzanian imports during the period increased more than eightfold, those of Ethiopia almost threefold. Madagascar imported most: 15 times more cereals in 1978 than in 1969-71. Egypt's imports, in the 'other countries' group, rose more than fourfold.

A number of countries stand out from this general deficit situation, however: these were net exporters or countries whose imports declined sharply. One such was Kenya which remained a net exporter and whose exports increased even further by 1978. In the rest of southern and eastern Africa, imports showed a declining trend in Botswana, Malawi and Uganda. In the other sub-regions, this downward trend can be observed only in the Central African Republic. It remains to be established whether or not these falls were due to inadequate means to finance imports.

As for food imports in the region, what has just been noted with regard to cereals is even more marked for some other basic food products. Thus, it is estimated that by 1985 milk import requirements will increase by 6.4% per annum and those of meat by 9.8%.

Food imports are thus rising year by year. On the basis of an index of 100 for 1969-71, the volume index of food imports for the region was 121 in 1972-74, 147 in 1975-77 and 210 in 1978. In value, between 1969-71 and 1978, imports rose by 389%. It goes without saying that for most countries that have serious external payments difficulties, this situation poses serious problems for food security and for economic development in general. Compared to that, international food aid destined for the region constituted only an addition of extremely limited scope. Thus, for example, the volume of food aid sent to the region increased by 25% per annum between 1974-75 and 1977-78. But that represented only 15% of cereal imports in the period.

Table 9 shows that calorie requirements are decidedly not being met. For the whole of Africa, calorie intake as a percentage of requirements was only 93 in 1969-71 and 94 in 1975-77. Considering the various sub-regions, it is only in North Africa that the norms could be met and this happened after 1972-74.

Conclusion

In the preceding sections we have endeavoured to present in figures a very worrying situation: the significant backwardness of African agriculture compared to that of other regions of the world and, above all, the growing difficulty that this agriculture experiences in being able to feed the peoples of the continent.

Except for a few mineral- or oil-producing countries, in general development policy in Africa has consisted, and consists, in extracting resources from agriculture in order to finance the rest of the economy. This has meant that in the agricultural sector, efforts were for a long time concentrated solely on products for which there is a demand on the world market. From the colonial period down to the present day, this policy has led to a continual decline in the capacity to produce local foodstuffs in order to satisfy food requirements. The consequences of the strategies and policies that underpinned these development options reached their critical threshold in the early 1970s. The great famines that struck much of the region at that time simply revealed explosively a crisis situation that had in fact existed, at

Table 9
Africa. Daily per caput calorie supply as percentage of requirements[1]

	Average 1969-71	*Average 1972-74*	*Average 1975-77*
North Africa:	89	101	105
Algeria	79	89	98
Sahel:	89	82	87
Chad	88	76	75
West Africa:	97	95	95
Guinea	89	85	83
Central Africa:	95	96	99
Angola	85	86	88
East and Southern Africa:	94	93	94
Mozambique	88	85	82
Total Africa region	*93*	*93*	*94*
Other countries[2]:	102	105	108
Somalia	96	96	92

[1] National food requirements are expressed in calories and are based on the approximate energy needs of working adults, taking account of height, age distribution and ecological factors.

[2] Invited members of the UN Economic Commission for Africa but not belonging to the regional conference.

least latently, since the first contacts, or more accurately the first clashes, between capitalist systems and African traditional systems, essentially based on subsistence.

In order to understand properly the nature of the crisis which is today assuming tragic dimensions, it is necessary to go back to the precolonial socio-economic formations to try and understand the nature of the modes of production that developed there, how the capitalist mode of production acted on those precapitalist modes, the forms of extortion of surplus peasant labour that that gave rise to, the social stratifications and class relations that flowed from all this, and the economic basis on which these social relations were consolidated.

This is the only way to understand the negative evolution, occurring before our very eyes, of the economic situation of most African countries and also be able, if not to draw up programmes for alternative policies, at least to reflect on the conditions and possibilities of these alternatives.

Notes

1. H. Mendras and Y. Tavernier (eds.), *Terre, paysans et politique*, 3 vols. (Paris, Futurible S.E.D.E.I.S., 1969).
2. Ibid., pp. 256 and 257.
3. This process is well explained by Suzanne Berger in *Terre, paysans et politique*.
4. *Studies of Trends in World Supply and Demand of Major Agricultural Commodities* (Paris, OECD, 1976).

2: Precolonial African Societies

Much work has been done on precolonial African societies in an attempt to determine what modes of production developed there, what types of contradictions marked their evolution and what were their particular characteristics compared to the classic schemas of a certain conception of Marxism. Heated debates have led to the confrontation of various theses. Some maintain that Africa did not escape the general laws of the evolution of societies which proceed from 'orthodox' Marxist analysis. Thus it is said that, apart from a few non-determining specific features, the same types of evolution from the primitive commune to feudal society, including slave society, are to be found there, as was the case in Western European societies. Other analyses try to show that many regions of Africa were marked by modes of production belonging to the 'Asiatic' type of societies pointed out by Marx.

Some students have tried to show that African societies do not fall into any of these well known classic categories current in Marxist literature, and that while respecting the method of approach of historical materialism, analysis of these societies would lead to the discovery of such particular characteristics that it is possible to speak only of a strictly African mode of production.

This attempt to understand the nature of precolonial African societies is not simply of speculative intellectual value. Indeed, it is only knowledge of this past—despite the tentative nature of the conclusions that might be reached, and uncertainty and inadequacy of the archaeological materials and the accumulated sociological data—which can throw light on the history of these societies and make it possible to correctly assess the problems relating to today's struggles, and bring out possible prospects for the future. Better understanding of the earlier societies will make it possible to assess the importance of the resistance to penetration of the capitalist system into the traditional sectors, and to situate the causes of the setbacks to current attempts at modernization in these sectors.

On at least one point all researchers are agreed: precolonial Africa had generally everywhere progressed beyond the stage of primitive communism. Bands of individuals grouped together and lived by hunting and gathering, sharing the common booty or common finds equitably. The very uncertainty of the condition of life necessitated total solidarity resting on an almost total egalitarianism.

If primitive communism had disappeared, then what type of society replaced it? Researchers are divided on the answer to this question. There are many arguments to defend each position, since knowledge of the subject continues to be uncertain and inadequate. Our aim then is not to reveal the truth which others failed to reach, but to attempt a synthesis that might contribute to understanding the stagnation, regression and crises that Africa is today experiencing, which are situated on various levels, political, economic and social.

Stalin, it may be remembered, listed five stages for the known evolution of the history of humanity. Given the very dominant influence of Stalin and the USSR on Marxist thought, and not only the thought but also the whole Marxist practice of the period, this statement was systematized into a catechism. Thus, a schema of the evolution of humanity in five stages on the example of Western Europe was established for the whole of humanity. It was not until the end of this period during which Marxist thought remained frozen, something which is contrary to its nature, that researchers began to ask themselves again about the real nature of earlier non-European societies (Asia and Africa notably) and research resumed leaving aside any schematic dogmatism.

People began again to study the writings of Marx and Engels on the question, as well as the often fragmentary information provided by historians, sociologists, anthropologists, etc. Thus, it was discovered or rediscovered that Marx had detected in the ancient East a quite particular mode of production which he had called the Asiatic mode of production.

It is worth recalling that the Soviet supporters at the Tiflis and Leningrad conferences (1929–31) had rejected this mode of production because of the implications of stagnation that it involved and which might lend credence to the idea that the socialist revolution was impossible in China.

Marx had not examined this mode of production in depth, and after the Stalinist period efforts were made to fill the gap. Later, numerous authors reached the conclusion that this mode of production was typical, in different forms, of both the societies of the East and Far East and African societies.

Numerous publications have appeared, either confirming or rejecting this thesis. In this framework, studies published under the auspices of the Centre d'Etudes et de Recherches Marxistes—Editions Sociales (C.E.R.M.), notably on precapitalist societies and the Asiatic mode of production, played a pioneering role. In the publication entitled *Sur les Sociétés Précapitalistes*, Maurice Godelier endeavoured to reconstruct what Marx actually thought using selected texts by Marx, Engels and Lenin, some of which had long been unknown to many researchers, a fact which had led to distortions and deviations throughout a whole historical period. Thus, for example, in order to show that the five-stage schema was never part of Marx's thinking, he quotes a letter from Marx, addressed to the editorial board of a publishing house in late 1877. Marx then wrote:

> But that is too little for my critic. He feels he absolutely must metamorphose my historical sketch of the genesis of capitalism in Western Europe into a historico-philosophical theory of the general path every people is fated to tread, whatever the historical circumstances in which it finds itself, in order that it may ultimately arrive at the form of economy which ensures, together with the greatest expansion of the productive powers of social labour, the most complete development of man. But I beg his pardon. He is both honouring and shaming me too much.[1]

But Maurice Godelier was not satisfied simply to try and re-establish Marx's true thought. He also expressed reservations not on the movement critical of the Stalinist period that began after the 20th Congress of the Communist Party of the USSR, but on the fact that this movement of criticism was at one on seeking a simple return to Marx, which in his view threatened to lead to new dogmatisms. According to him, while keeping in mind this heritage, it was necessary to take account of the much more advanced state of information and scientific knowledge now available about precapitalist societies, in order to purify this heritage and eliminate the 'dead wood' from it.

On the subject of the Asiatic mode of production, he considered as 'dead wood' the notion of 'oriental despotism' and the idea of 'unchangeableness' advanced among the typical characteristics of this mode of production. We shall have occasion to return to these issues in the course of our reflexions on precolonial African societies.

In the C.E.R.M.'s second book devoted to research on the Asiatic mode of production, Jean Suret-Canale concluded that there did exist some forms of this mode of production in some ancient African societies. In the same book, Maurice Godelier moved in the same direction while Jean Chesneaux raised questions, and Catherine Coquery-Vidrovitch thought rather that research should be directed at discovering a peculiarly African mode of production, since she was unable to find in earlier African societies the basic features typical of the Asiatic mode of production. There are, however, still many authors who accept the five-stage schema of the Stalinist period, especially among Soviet authors and researchers and those who think like them. They generally place ancient African societies in the slave system or in the little developed stages of the feudal system. Majhmout Diop in his study, *Histoire des Classes Sociales en Afrique de l'Ouest*, adopts this approach.

We shall endeavour to examine the main theses advocated in order to have at least an outline, inevitably no more than a tentative one, of the nature of precolonial African societies.

In his study 'Les sociétés traditionelles en Afrique tropicale et le concept de mode de production asiatique', Jean Suret-Canale thinks that many precolonial African societies belonged in the family of the Asiatic mode of production.[2] Broadly speaking, he distinguishes two main types among tropical African societies existing at that time: tribal or tribo-patriarchal societies and class societies. The former he sees simply as societies in transition between the primitive community and class societies. As a result the possibility appears of forming a surplus and hence too the possibility of exploitation. It seems that these societies persisted above all in regions with low agricultural productivity, as was the case, for example, in the dense forest zones where difficulties in clearing land considerably reduced productivity. In addition, the very nature of the products (plantation crops more difficult to conserve) did not make it easy to accumulate a surplus.

It followed that the need for the exploitation of man by man was less marked, and society tended to remain more egalitarian. The families of the first occupants of the land would levy tributes from other incoming families which would form the village. But, in general, these tributes were only compensations for the responsibilities that had fallen upon them, religious responsibilities or responsibilities for distributing land. Even in the event of war, the captured enemy was simply incorporated into the family where he contributed to the common production. But the seeds of class exploitation existed already since the chiefs of the land (the first occupants) who were sometimes also war chiefs could, by conquest, extend their rule and domination over other villages from which they would also levy tribute. That could lead to the exploitation of a whole set of village communities opening the way to the appearance of true exploiting classes.

In short, these were stateless societies, because in them classes were not truly formed—in which the surplus product and productivity were extremely low and in which, therefore, the level of productive forces was still very low and the division of labour still embryonic. The consequence of all this was that there was little exploitation of man by man, with even prisoners of war not being true slaves, that is men who had the right to live only in order to work for others. This type of social organization is said to have existed among the Fang in Cameroon, the Manssangou in Gabon, the Kissi in Guinea, etc.

In his analysis of class societies, Jean Suret-Canale stresses that

> without falling once again into an excessive geographical determinism, it is quite clear that the savanna environment with its cereal agriculture and the necessity, imposed by the seasonal rhythm of the climate, to build up reserves with the facilities of circulation, trade and diffusion of techniques that it involves, lent itself particularly well to the deepening of social conflicts.

Quoting Jean-Jacques Maquet, he adds:

> the surplus product, the surplus, was particularly significant there because it could be made up of cereals and legumes. They can be preserved almost indefinitely, are easy to transport and easy to measure and are uniform enough to be comparable. These characteristics make possible a certain accumulation of a very movable wealth. Thus, the surplus can be easily separated from the producer, pass from hand to hand and be concentrated in a relatively large quantity.

When therefore the conditions of accumulation of the surplus improved, there emerged a tendency to make the defeated into slaves working for patriarchal families.

The war chiefs and the chiefs of the land then began to embark on raids and wars to extend their power over families and villages, and this became a means of enrichment. Slavery tended to become universal with the constitution of slave families and slave villages and hence of antagonistic classes. Must we then conclude that a slave society existed? Jean Suret-Canale thinks not. Slaves were exploited through being obliged to perform services without them having any well-defined role in the masters' system of production. On this point he notes:

> The large agricultural estate based on servile labour or artisanal workshops based on slave labour were always unknown in Africa or only appeared there exceptionally...The condition of captive, although very widespread in Africa, had an essentially legal content and did not imply any given role in the production which characterizes a social class.

Jean Suret-Canale believes even less in the existence of a feudal society, since the relations of production which characterize this mode of production are based on a landed hierarchy, the feudal lords having private and personal rights over land. African societies which did not have private ownership of land could not be included in this framework.

The use of the term feudalism in this case could at most have a political meaning. The contradictory development of class society was to lead to the appearance of the state in its role as preserver of the interests of the dominant classes. The African states would be consolidated thanks to their economic role based essentially on trade, and particularly long-distance trade.

In conclusion, Jean Suret-Canale proposes a sufficiently broad definition of the Asiatic mode of production that it includes African societies:

> It remains that, in our opinion, the socio-economic structure characteristic of the Asiatic mode of production is limited to the coexistence of a production apparatus based on the rural community, the collective ownership of the land to the exclusion of any form of private ownership, and the exploitation of man by man in forms which may be extremely varied, but which always pass through the intermediary of the communities.[3]

That, of course, pushes into the background the problem of irrigation which had struck Marx as the economic role of the state in Asiatic societies. In reality, for Marx this problem of irrigation was never a determining feature of the characteristics of the Asiatic mode of production. As Jean Chesneaux,[4] Maurice Godelier,[5] Jean Suret-Canale and many other Marxists have shown, this question of irrigation was only emphasized by well-known anti-Marxists, inspired by the determination to ridicule Marx's idea and discredit the socialist world. Karl Wittfogel[6] stands out in this respect. Taking the notion of the Asiatic mode of production, he reduced it to a 'hydraulic society' whereas for Marx, irrigation works such as the great unproductive prestige works were only possible bases for the appearance of a state dominating the primitive communities. Thus, according to Jean Suret-Canale, Africa experienced a form of the Asiatic mode of production with trade, and especially long-distance trade, playing the economic role in consolidating the state.

Catherine Coquery-Vidrovitch in 'Recherches sur un Mode de Production Africain'[7] disputes the main points put forward by Jean Suret-Canale that we have just examined. First, she thinks excessive importance is given to the notion of state and stateless societies in Africa; next she rejects 'the definition of a surplus value exclusively based on the diversion of the labour of villagers by privileged strata'; finally, she finds the definition of the Asiatic mode of production given by Jean Suret-Canale 'too general to be operational.' At any event, she thinks that the earlier African societies evolved in ways totally different from the ways in which both European and Asiatic societies evolved. Research should thus be directed towards defining a strictly African mode of production. As for the basic difference between the types of evolution observed in Africa and Europe, there does not seem to be any serious disagreement between her and Jean Suret-Canale who also accepts it.

Her analysis is thus centred on demonstrating that the same basic difference exists between the types of evolution that developed in Africa and in Asia. She starts from a narrower definition of the Asiatic mode of production which 'presupposes on the one hand village communities based on collective productive activity' but bound to a 'higher unit' which, in the form of a state regime, is capable of compelling the mass of the population to work collectively; this 'generalized slavery' reveals the 'economic high command' of a despot who 'exploits the communities at the same time as he rules them'.

According to Vidrovitch, the dynamic element, that is, the motor of the evolution of society, in this mode of production remains the massive public works which the village communities are forced to execute by a despotic government. This she finds nowhere in Africa. But rather 'the economic life of precolonial African societies was characterized by the juxtaposition of two apparently contradictory levels: on the one hand village subsistence, on the other, big international, even intercontinental trade.'

With numerous rich historical references she endeavours to show the basic importance of long-distance trade in the evolution of a whole range of African societies. On top of that is the fact that, even in the most centralized societies, the tribo-patriarchal structures remained the form of village organization; she therefore feels it necessary in Africa to get beyond 'the traditional contrast between state society and stateless society.' In addition, according to her, the major exactions, that is, the most intense exploitation was not exercised on the subject village communities; 'they came from outside the territory thanks to annual raids or to peaceful commercial transactions.' 'The specificity of the African mode of production,' if we understand correctly, the motor of the evolution of African societies, is seen as resting in 'the combination of a patriarchal economy and the exclusive ascendancy of one group over long-distance trade.'

Thus, Catherine Coquery-Vidrovitch concludes that the African mode of production is not reducible to the precapitalist modes of production in the West and the Asiatic mode of production 'in the absence of a true despotism aiming at the direct exploitation of the peasant class'.

As for Majhmout Diop, in *Histoire des Classes Sociales en Afrique de l'Ouest* he puts himself firmly in the line of classical thinkers who consider that Africa evolved 'like all human societies except for a few minor variations and that it had the primitive community, slavery and if not full feudalism at least its lower forms.' His argument is based mainly on the large numbers of slaves and their integration into the system of production. Not only was the mass of slaves very large, but also only a tiny fraction was considered as domestic slaves, that is, as servants. The others 'slaved away in the fields', that is, they were integrated into the system of production. He illustrates his analysis by statistics taken from Mali and Songhai.

Majhmout Diop concludes the chapter on the 'nature of precolonial society' in these words:

> The originality of our society both as regards the castes and in the organization of village communes and collective ownership of the land scarcely goes beyond the framework long ago outlined by Marx, who had elsewhere foreseen the wealth of possible forms and stages. There is, therefore, no need, it seems to me, to seek specific modes of production.

Since we do not intend here to offer anything new with regard to such a crucial problem which would require years of interdisciplinary research, we shall aim for a degree of simplification which might make it possible to bring out the main features of precolonial African societies.

Thus, we have thought to limit to these three main theses those that have attempted to define these societies, given that each of the theses covers a multitude of variants.[8] What is particularly interesting in these three theses is that their main exponents, Jean Suret-Canale, Catherine Coquery-Vidrovitch and Majhmout Diop, looked especially at the case of Africa in the firm hope of discovering the main bases of their theses there.

We shall now compare the three theses to bring out those features on which they seem to be agreed and to uncover the contradictions between them. This will help to pinpoint the structures that seem essential in the socio-economic formations that precolonial Africa experienced, and might contribute to throwing some light on the evolution of contemporary Africa.

A first key point, and one which seems indisputable, is that the basic social organization in these societies was in general the village community with collective ownership of the land. A second important point is the recognition of the existence of slavery in most of these societies. The divergences seem to lie only in the assessment of the type of exploitation to which the slaves were subjected and of their role in the system of production.

For Majhmout Diop, it is enough to observe the number of slaves involved in the system of production, even simply at the level of family lineage, to conclude that a slave-type society existed.

Jean Suret-Canale responds to that conclusion that, so long as the conditions of their incorporation into the system of production do not differ from those of other members of the lineage, slavery cannot be said to exist, the slaves simply contributing an extra labour force to the extended family. According to him, such a conclusion would require in addition large agricultural estates based on slave labour, which did not exist anywhere in Africa.

Catherine Coquery-Vidrovitch also recognizes the existence of slaves but thinks that the most significant exactions operated by the aristocracies did not arise from the exploitation of the subject peoples in the territorial framework, but came from outside, through raids or peaceful international trade.

In short, it can be said that slavery did exist in precolonial African societies but in attenuated forms—no large-scale public works being especially set aside for the subject masses—the aristocracies contenting themselves for the most part with levying tributes from the subject peasant collectivities. This would constitute a clear distinction from other societies in which despots subjected the enslaved masses to special labours in which noble or aristocratic families took no part. Because of this, the confusion with societies that were based on slavery is not sufficiently explained.

But other key problems remain and they particularly separate Jean Suret-Canale and Catherine Coquery-Vidrovitch. First there is the problem of the state, and second the role of long-distance trade.

For Catherine Coquery-Vidrovitch the distinction between state societies, and stateless societies based on the existence of a considerable surplus that is exchanged and leads to a struggle over this surplus, the appearance of classes and consequently of a state power that succeeds in seizing the surplus, seems artificial and not one that can be systematized for Africa. Referring to Georges Balandier who stresses the heterogeneity of precolonial African societies, she insists on the imbalances that can arise in a society based on 'a tribal and lineage structure based on the family and a territorial organization with a more or less centralizing tendency' without that leading to distinctions between stateless societies—in which trade would be virtually non-existent—and state societies—where trade would constitute vast networks. We are not claiming to settle such an important question so easily.

It seems to us, however, that whatever the heterogeneity of ancient African societies, the key thing appears to be to know whether with the development of agriculture—the primitive commune based on hunting and gathering having for the most part disappeared—the appearance of a significant surplus, whose acquisition would lead to intense struggles possibly involving the appearance of classes and hence of the state, did or did not occur. If it did happen, as we believe, then forms of state existed in most African societies, superimposing themselves over kinship relations of domination and dominating the lineage families and village communities.

Are not the imbalances, to which Coquery-Vidrovitch refers, 'between a tribal and lineage structure based on the family and a territorial organization with a more or less centralizing tendency' tied up with the appearance of social classes becoming detached from the rest of society, dominating it and organizing it in a certain state form? For ultimately, cannot one consider as a state a form of social organization in which one social class—Coquery-Vidrovitch speaks of an aristocracy—dominates and exploits the rest of society whether or not organized into distinct classes? If such is indeed the case, it can be said that the precolonial African kingdoms and empires were states whose nature would of course need to be defined. In support of this thesis, Maurice Godelier maintains that 'in order to find a fully developed state structure, one must leave the Polynesian area and turn to the traditional African states or the states and empires of pre-Columbian America'.[9]

Finally, looking at most of the authors who have been especially concerned with ancient African societies, one is inclined to think that structured and organized state forms did exist in many societies. Moreover, Catherine Coquery-Vidrovitch does not deny it since she herself cites the empires of Ghana, Mali, Benin, etc. The nature of these state forms may be similar to the Asiatic mode of production on condition that its geographical character is removed and it is not too narrowly defined. Indeed, some definitions do not reflect the thought of Marx himself, such as the problem of 'hydraulics' as the sole public work activity imposed on the subordinated communities.

As for the notion of 'despotism' that Catherine Coquery-Vidrovitch uses in the definition of the Asiatic mode of production, does it not refer to the appearance rather than to the reality of power? The rulers of African states had very wide powers but they did not always exercise them oppressively as the notion of despotism would tend to suggest.

However, Maurice Godelier mentions cases, even in Asia or other regions, where this notion of 'oriental despotism' is more apparent than real. He mentions as examples the power of the emperors of China and the government of the Incas. Here too, then, one is tempted to think that the power of the rulers of Africa and those of Asia was often very similar. But there remains the important question of long-distance trade.

For Jean Suret-Canale commercial exchanges played only a consolidating role in African states, states born of the sharpening of class contradictions internal to these societies and their mode of production.

Catherine Coquery-Vidrovitch defines the mode of production that Africa experienced as 'the combination of a patriarchal communal economy and the exclusive ascendancy of one group over long-distance trade.'

For our part, it seems difficult to conceive external trade as a basic feature in the process of formation of a mode of production. It seems to us that long-distance trade, by bringing in external factors, could only act to strengthen or weaken an existing mode of production already formed from social antagonisms internal to the society.

Whatever the importance of unequal exchange, discussed by Catherine Coquery-Vidrovitch, which operated to the advantage of the aristocracies—'peaceful commercial transactions which ensured the monopolizing of products at rates lower than their value'—it seems to us that that could at most act to consolidate or weaken the internal relations of production without being responsible for the formation of those relations. One may even wonder about the origin of the products exchanged by the aristocracies with the outside world. Did they not come from the accumulation of the surplus extracted from the exploited village communities? If such is the case, why were the aristocracies not more predisposed to increase the exploitation of these communities in order to ensure for themselves even more foreign goods through long-distance trade?

Finally, one last question, and not the least important one, concerns the forms and possibilities of evolution of precolonial African societies, whether they are considered as partaking of the Asiatic type or are considered as specifically African. Much stress has been laid on the stagnation of societies possessing the characteristics of the Asiatic type. Maurice Godelier maintains the opposite was the case, and with historical examples endeavours to demonstrate rather that they were progressive societies.

Whatever the case, while some societies comparable in many ways to the Asiatic societies experienced stagnation, this stagnation may be explained by the resistance to the explosion of the principal contradiction peculiar to the mode of production involved which is 'class exploitation and maintenance of collective ownership of land'. This resistance itself may probably be explained by the fact that exploitation tended to strengthen the communal structures and the collective ownership of the land and that these same structures facilitated exploitation. The resolution of the contradiction in a progressive direction could only arise from the break-up of the collective ownership of land through privatization of the land following, for example, the development of private trade.

As for the African mode of production as it is put forward by Catherine Coquery-Vidrovitch, she writes of it:

> The occasions for a blockage are certainly more frequent than elsewhere since the productive forces are not true forces based on war or trade, production is sterile, the surplus is certainly assured for the privileged class but it is an apparent surplus, the ransom for which, in the longer or shorter run, is the impoverishment of the country.

This real impoverishment involving regression or stagnation would come about as soon as trade flows move in other directions. It would seem, then, that the tendencies to stagnation or even regression were strong in the earlier societies that existed in Africa. It is striking to observe that even the importance of trade and its volume which constitute the basis of Catherine Coquery-Vidrovitch's analysis failed to destroy the system of collective ownership of the land. This virtual absence of a property regime based on private ownership until colonization, and despite the development of trade and a certain amount of accumulation, is doubtless to be explained in part by the relative abundance of land and the rudimentary level of the material means of production.

In conclusion, we shall not try to define the nature of the modes of production that precolonial sub-Saharan Africa experienced. We have simply tried to sketch in the outlines from the work of specialists. More detailed multidisciplinary work still needs to be done to provide more detail on the nature of the precolonial African modes of production and social formations.

However, it is possible to say here that rigid communal structures at the level of village communities dominated and exploited by aristocracies caring more for wealth and display than for the development of productive forces, ruling states that were sometimes politically well organized but with a fragile economic base, meant that precolonial Africa had numerous characteristics that might lead to stagnation or even regression. It was regression which was unleashed by the slave trade that brought large-scale destruction of the productive forces and later facilitated colonial penetration.

Notes

1. M. Godelier, *Sur les Sociétés Précapitalistes* (Paris, Editions Sociales, 1970), p. 14 (Eng. tr. in L.S. Feuer (ed.) *Basic Writings on Politics and Philosophy: Karl Marx and Friedrich Engels* (Garden City, 1959), pp. 440-1).
2. *Sur le Mode de Production Asiatique* (Paris, Editions Sociales, 1970).
3. Ibid.
4. Ibid.
5. Ibid.
6. *Oriental Despotism: A Comparative Study of Total Power*
7. *Sur le Mode de Production Asiatique*, p. 345.
8. Samir Amin's studies of the tribute-paying mode of production in *Unequal Development* (Hassocks, Harvester Press, 1970) are of great interest as are the criticisms by Amady in 'Formations sociales et commerce à longue distance'.
9. Maurice Godelier, *Sur les Sociétés Précapitalistes*.

3: The Appropriation of Peasant Surplus Labour

If the modes of production that existed in precolonial Africa had, as we have seen, strong tendencies to stagnation or even regression (and these tendencies were accentuated by the slave trade), the African societies of that time that rested on these modes of production could not prevent the settlement of colonial invaders despite often very strong resistance. As they installed themselves so capitalism reduced these precapitalist modes of production to its domination.

In what form, and through what types of relations was this subordination secured? It should be recalled that the beginnings of the colonial era were preceded by the establishment of merchant capitalism along the African coasts in the form notably of trading posts. This merchant capital performed the role of a transmission belt marketing tropical products in the metropoles in order to supply the holders of industrial capital. In the metropoles themselves, the integration of industrial capital and finance capital was underway but was still just beginning.

Torn between its concern to secure protection for itself against African rulers and its fear of seeing the metropolis penetrate and administer the African continent directly—which might challenge its role and monopoly as middleman—this overseas merchant capital was soon overtaken by the competing interests of different European nations. Under the sway of more powerful industrial and finance capitalism in Europe, and anxious to carve out for themselves protected economic spaces overseas which would be exclusively reserved to them, the nations of Europe embarked on colonial conquests, each one trying to seize, control and administer as much territory as possible. In this way, European national industrial capital was able to ensure for itself sources of raw materials and outlets for its products, well sheltered from international competition. In the same way, it was thought that finance capital could hope for new investment possibilities for this capital.

But reality was very different. For much of the colonial era, industrial and finance capital played only an insignificant role in the exploitation of the African colonies. On this point Catherine Coquery-Vidrovitch notes: 'Neither did the partition of tropical Africa, carried out with no major pre-existing economic motive, give rise to the development that might have been expected. Its exploitation was a late display of archaeo-imperialism. Until the Second

World War the colonial trade economy there continued to be based on pillage.'[1] This makes it possible to spell out the nature of the capitalism to which the precapitalist African social formations were subordinated: capitalism of an essentially merchant and archaic character with no concern to invest and even incapable of doing so, animated by the feverish desire to extract from African productive systems as much as possible, as quickly as possible and as cheaply as possible.

It was the objectives, the means and the methods of this archaic capitalism, which were to play a leading role in the nature and scale of the changes that the pre-existing modes of production experienced, which were examined in the previous chapter.

If capitalism in general is preoccupied by the quest for profit, merchant capitalism was here concerned with the maximum appropriation of the surplus labour of African manpower. Its behaviour towards the relations of production already existing in these societies was a function of these concerns.

So long as the defeated aristocracies were disposed to co-operate so as to extract more from the surplus labour of the peasant masses, it did not seem necessary—indeed quite the opposite—to tamper with the pre-existing social hierarchies. Thus, when the traditional chiefs were agreeable, they were kept in office. Only the recalcitrant were exiled. To achieve greater efficiency in the exploitation of the peasant masses, a colonial administration was added on top of the African aristocracy.

Depending on the circumstances and the places, however, the modes of production sometimes suffered major changes, carried out of course by means of the harshest violence. This was what usually happened when capitalism deemed it in its interests: the lands of the peasants were seized; labour was removed from areas with a low productive potential to areas with a high potential; money was introduced in order to impose a certain type of production.

These phenomena did not all have the same importance everywhere. But they played their part in modifying some rules and customs of the traditional social formations:

1. The expropriation of the peasants led in some regions to the appearance of private ownership of land and hence destroyed, partly at least, its communal appropriation.
2. The forced removal of labour led in those areas where this had taken place to partial dislocation of the lineage.
3. The introduction of money for the payment of taxes, in particular, forced the traditional formations to enter commodity systems.

All that is reminiscent of the primitive accumulation of capital described by Marx in *Capital* and which relates to the period in the West in which capitalism was being consolidated on the ruins of feudalism. There are, however, considerable differences.

In the first place, except for the problem of the integration of the precolonial formations into the commodity system through the introduction of money, the two other phenomena mentioned above—the dispossession of the peasants and removal of labour—did not occur all over the continent, but here and there. Hence, the objective of capitalism here was not to dissolve traditional modes of production. So long as these could be useful to it, capitalism saw no need to do so. In the West, on the other hand, the objective of capitalism was in general the dissolution of the feudal system.

Another major difference relates to the means used. Whereas in the West the dispossession of the peasants and the constitution of a labour market in the service of capital were essentially achieved through the economic power of capitalism, in Africa the expropriation of the peasants and the removal of labour were achieved by means of brute force. In the last analysis, it can be said that colonial capitalism accommodated itself to the African socio-economic formations when they served it well and sometimes altered them considerably when its interests so demanded.

Thus, according to the productive potential of the subject regions, it can be observed that the colonial system operated two broad types to establish its domination:

Colonies of settlement in which the colonial administration dispossessed the indigenous peasants of their lands to the benefit of European populations. This was the case in North Africa, notably in Algeria, and in much of southern and eastern Africa.

Colonies of exploitation in which the colonial administration made no particular effort to settle the land with white settlers, but rather sought to have the land worked by the peasants for the benefit of the metropolis. This was the case in much of Africa south of the Sahara.

While the forms of exploitation varied, the means used and the objective aimed at were quite similar. It involved using methods of force and forced labour to exploit the labour force massively while avoiding recourse to machines which, produced in the metropolis under the capitalist mode of production, were expensive and raised the costs of production.

The forms of exploitation in the settler colonies were more reminiscent of the forms of the primitive accumulation of capital which had developed in the centre at the beginning of the industrial era. These were the violent dispossession of the peasants, the granting of their lands to white settlers, the transformation of the dispossessed peasants into landless peasants obliged to work in the plantations acquired by the settlers, or in the mines where underground resources were abundant.

In the settler colonies of British-ruled southern Africa, the populations were driven back and concentrated on small poor lands thus constituting inexhaustible labour reserves. Since the conditions of production did not allow them the means to survive in these lands, these populations were reduced to working as wage labourers in the settlers' plantations or to being taken on as workers in the mines. Thus proletarianized, they constituted only a bastardized proletariat to the extent that the bosses in no way assured them the full conditions of the reproduction of their labour power: their conditions were very low wages, no social security, no old age pensions.

This manner of treating wage-earners was indeed typical of the whole colonial system in general. The costs of maintaining wage-earners were never wholly paid for by the system, wage-earners having no choice but to rely on the peasant, family-based economy to guarantee them a significant portion of the means of reproduction of the labour force placed at the service of capital. The wage was in fact simply a little topping-up, and through the wage-earners it was the peasantry who suffered an extra form of exploitation.

The enclosing of the peasants of southern Africa in reservation areas and the possibility for capital of obtaining however much cheap manpower it needed fully justified describing this part of the continent as the economy of the labour reserves.

Elsewhere, in tropical Africa, the establishment of white settlers on the land was rather limited. The colonial system organized the trading economy based on the production of tropical products in exchange for manufactured goods. But what should be emphasized is that this was not a free exchange. The peasants were not free to produce what they wanted, but only to produce what the colonizer wanted and on the conditions determined by him. For that every form of constraint was used.

In the beginning the colonial trading economy rested mainly on the products which could be found on the spot: palm products, timber, shea butter, jute, gum and other products of gathering. Groundnuts and cotton were developed later and were followed by the introduction of new crops previously unknown in much of Africa. Even in 1914, products of gathering still accounted for half the exports of French-ruled Africa. From this period the new products began to take over first place. Throughout the first period the most archaic means—which were also almost without cost—were used to ensure production.

The system of crops and forced labour which remained the general rule led to mediocre results at the level of production which remained unchanged or fell year by year despite the increased resort to force.

It was with the development of cash crops, which offered great possibilities for profits, and in order to increase yields that recourse was had to capital. Thus people began to buy the peasant's crop directly from the peasant, naturally at a price below its value. Compared to the conditions that had prevailed previously and which meant that the village chief was paid a fixed sum for the entire production of the village, this was already progress. The peasants generally saw none of this payment which the village chiefs and their entourage tended to appropriate.

The spread of the use of money and the increase of direct trade with the peasants led to the reduction of compulsory crops; this made it possible to relaunch production, especially of new crops. Thus in French West Africa coffee production rose from 60 tons in 1920-24 to 11,101 tons in 1935-39. Cocoa production rose from 2,586 tons in 1920-24 to 49,871 tons in 1935-39.[2]

While the colonial companies increasingly resigned themselves to purchasing from the peasants, producer prices remained excessively low. In 1925, for example, the colonial companies purchased a kilo of cottonseed from the producer at 1.25 F, and sold cotton lint at 12.50 F in the metropolis. If processing and equivalent quantities of cottonseed and cotton lint are taken into account, they were making at least a 300% profit. These companies deemed the price paid to the peasants too high, however, and persuaded the colonial administration to reduce the price; it fell to 0.60 F/kg. in 1933-34.

The development of the colonial trading economy which consisted in using small amounts of capital to remunerate the peasant producers was not accompanied by a parallel investment in machinery to improve production techniques.

At the level of cultivation of the land, ploughing, sowing, harvesting, maintenance work on the plantations, the initial processing of products (decortication, ginning), in short all the operations, were done manually. And this was in spite of the establishment of some settlers on plantation lands in countries which were not particularly suited to being settler colonies. On these settler lands the same archaic methods were employed by a labour force doing forced labour and later, paid labour. Until the post-war period, the only novelties to appear in the production of the colonial trading economy as compared to the precolonial period were related simply to the nature of the products, with the introduction of cash crops. The colonial trading economy thus relied entirely on the outright exploitation of the peasant labour force: from cultivation to initial processing, from the transport of products to the buying centres, to storing and loading, everything rested on physical labour.

During the same period, food was in increasingly short supply because of the decline of food crops in favour of export crops. The dramatic consequences involved a decline in population. Some regions became depopulated as a result of people fleeing to areas with a secure food supply. Sometimes it was the colonial administration that moved some peoples from regions with a poor agricultural export potential to regions where there were serious labour shortages.

Ultimately, until the post-war period, the colonial economy was simply an all-devouring pillage economy organized by the colonial companies and enterprises, and the demands of rapid accumulation had a much higher priority than any concern for rational exploitation. The more the super-profits realized in the colonies accumulated in the metropolis, the more the destructive practice of pillage developed.

In fact, the use of archaic and irrational methods in the exploitation of the resources of the colonies corresponded to the establishment of a capitalism that was itself backward and little developed. One-man enterprises with limited means were the most common form. In 1956, these were still responsible for 61% of the business in French West Africa, as against 39% for the private sector companies.

This type of capitalism could only survive in the colonies because, in the metropolis, capitalism was still more national than international in character.

The formation of an economically united Europe with common customs, the result of the growing internationalization of capital, especially industrial and finance capital, was to put an end to the reign of purely national companies in the centre. There necessarily ensued a review of the privileged and protected situations enjoyed by one-man enterprises in the colonies. With the growing internationalization of capitalism, larger financial means were made available to develop exports from the colonies. In the same way, the form of political domination was to be profoundly altered. Direct domination was gradually to give way to administrative autonomy.

Industrial and finance capital came increasingly to occupy the mining and industrial sector. The old colonial capitalism was more or less sacrificed and some economic positions were abandoned to local privileged strata living in the orbit of colonialism: planters, traders, officials in the administration. Thus, the process of forming an African bourgeoisie got underway. The birth of this bourgeoisie is linked to changes in capitalism which, in order to modernize these forms of exploitation of the colonies, needed class allies on the spot to launch the process of emancipating the colonies that was rendered necessary with the internationalization of capital. The contradictions between bourgeois fractions in the metropolis were exacerbated by this problem of emancipation. But the struggles waged by the colonized peoples themselves facilitated the triumph of the most modern and most powerful fractions of central capitalism, which were advocates of an emancipation which would inaugurate the neocolonial era and make possible large-scale exploitation at little cost (reduction in the costs of administration and the maintenance of order) to the colonial system.

These struggles against classic colonialism were led in the colonies by bourgeois or petty bourgeois fractions determined to strengthen their particular economic positions at the expense of small colonial enterprises, but in alliance with metropolitan big capital which also favoured modern forms of domination, leaving a considerable political and economic place to these local privileged strata.

These bourgeois and petty bourgeois fractions were active in a variety of areas. In plantation agriculture they waged a ferocious struggle against forced labour in order to secure labour on the same terms as the white settlers. The success achieved in this area enabled them to exploit cheap wage labour, generally in the plantations, the largest of which belonged to former traditional chiefs.

In seasonal savannah agriculture these privileged strata played above all the role of intermediaries between the peasant producers and the colonial companies, controlling in this way the collection and resale of peasant production and selling the peasants imported goods in return. This intermediary role enabled them to engage in large-scale speculation.

Among these privileged strata, there also figured prominently politically active intellectuals who were determined to give themselves an economic base by exploiting their positions in the neocolonial state apparatuses conceived as part of the new forms of domination prepared by big capital. These various social strata received their shares—crumbs, of course—of the surplus labour extorted from the peasant masses. They were to constitute the dependent bourgeoisies, appendages of big capital, and would be the dominant classes of the new states whose independence they would negotiate.

Thus, then, the manner in which independence was achieved—negotiation between, on the one hand, the most powerful fraction of metropolitan capitalism and, on the other hand, the African bourgeoisies and petty bourgeoisies which were consolidated through the exploitation of the peasants during the colonial period—broadly explains the way in which the domination of Africa, or more exactly the African masses, was considerably and rapidly increased after independence.

The fact is that the colonial system had become an obstacle to the development of the combined interests of the African privileged strata and Western big capital in an expansionist phase. With independence, imperialism removed all the barriers standing in the way of its expansion.

The intense struggles of the African masses were simply exploited for tactical ends by the African social classes and strata allied to imperialism to decide in favour of Western big capital. This capital had absolutely no desire to continue to bear the burden of colonial administration and take the risk of confrontation with the African masses which might lead to revolutions. And all this they did for and on behalf of the interests of a colonial capitalism that had become truly inappropriate in a world in which the imperialist monopolies were conquering all before them.

Thus the colonial system was rapidly liquidated in order to open up wider vistas for modern imperialism.

But far from bringing about an improvement in the conditions of existence of the essentially peasant African masses, this political change rather accelerated the crisis. This was so because the new African states, in accordance with the interests of the dominant classes, sought to go as far and as rapidly as possible to integrate their economies into the capitalist world market.

It is thus that, with the achievement of political independence, the extraversion of African economies was given a greater impulse.

Notes

1. C. Coquery-Vidrovitch in IDEP/Reproductions 1243: 'De l'impéralisme ancien à l'impéralisme moderne'.
2. Figures provided by C. Coquery-Vidrovitch in her study of Africa and the crisis of the 1930s; colloquium at the University of Paris VII, 9-10 April 1976.

4: The Export-oriented System

Problems of growth emerged as a matter of concern in the developed capitalist economies only after the Second World War. Previously, people were rather concerned with problems of general equilibrium between supply and demand and the optimum use of resources.

The problem of growth, which was posed as a goal and became a centre of interest from the 1950s, almost certainly because of concerns linked with post-war reconstruction, led increasingly to studies and research and came to be taken into account in economic policies.

This was all that was necessary for it to be thought, a decade later, in the 1960s, when most of the countries formerly dominated by colonialism became politically independent, that the solutions to the problems posed by the poverty and destitution affecting most of their populations would be found by promoting development policies centred on the rapid growth of the national product. The laws of the market were deemed sufficient to achieve a suitable redistribution of income.

The strategy thus proposed to the new African states was based on development through high growth of the national product. Efforts therefore needed to be directed towards reducing the constraints that might constitute obstacles to this growth. The central core of this strategy was to promote and ensure the expansion of development poles in which investment would be concentrated so that they would act as beacons for all the national economies. Investment, usually financed from external sources, was thus directed towards the relatively modernized sectors centred around primary export products (agriculture and mines). Of course, this was not an innovation, but simply the systematized pursuit of the economic policy that prevailed before independence, during the colonial period. It was argued that once this policy had been given national goals, it could naturally lead to the transformation of structures and to development.

To loosen the constraints of this chosen path, the new states had massive recourse to foreign capital for investment, as well as to foreign aid to ensure manpower training to meet the needs of the economy for skilled personnel.

The 'modern enclaves' which were to form the development poles could, in this perspective, only be the export sectors, that is, the primary products that represented the sole exportable resources of the new states. The export income thus accumulated would make it possible to finance development, notably through the purchase of machinery to begin some industrialization.

In agriculture—the economic mainstay of the vast majority of African states—the economic policy advocated took the form of excessive encouragement to the growing of export crops to the detriment of food crops. Table 5 (see Chapter 1), which shows the evolution of Africa's share in world production of the main crops of the region, brings out the expansion of export crops whereas crops intended for consumption were in decline.

Thus, between 1948-52 and 1971-72, Africa's share in world production of millet and sorghum fell from 17% to 13%; that of sweet potatoes and yams (other staple foodstuffs in the region) fell from 23% to 16%. Over the same period, the share in world production of coffee rose from 12.5% to 27%; that of cocoa from 65% to 72% and that of cotton from 9% to 11%. Between 1960 and 1970, the annual growth rate in the production of basic foodstuffs (cereals, root crops and legumes) was about 2.5%, whereas that of non-food products mainly intended for export (coffee, tobacco, cotton, rubber, etc.) was about 4%. Over this same period, the areas devoted to basic foodstuffs increased at an annual rate of 1.2%, whereas those devoted to export crops increased at an annual rate of 2%. In addition imported inputs were intended essentially for export crops. By way of example, 80% of the fertilizers consumed in Kenya were used on coffee and tea plantations. Uganda devoted 84% of its fertilizers to tea and sugar, and Senegal 52% of its fertilizers to groundnuts.

These figures, taken from the documents of the United Nations Economic Commission for Africa, bring out the growing interest in cash crops after independence. Devoting so much of Africa's meagre resources to these crops was bound to aggravate dangerously the food production situation.

This risk was taken in the hope, it was said, of developing domestic accumulation in order to finance imports of needed machinery. Ultimately, these objectives put forward to justify such a policy were almost never realized. The knock-on effect expected of the development poles did not take place. On the contrary, national economies rapidly became exhausted. Export income, which declined year by year because of the unequal system of international trade, was soon absorbed by the political and administrative bureaucracies and government prestige expenditure. In most countries, even the small industries established during this decade were rarely built from local resources.

This 'development' strategy, which virtually condemned food crop agriculture although the latter remained responsible for feeding the whole of a rapidly growing population, provoked considerable imbalances leading to the accelerated and massive impoverishment of the rural areas. Thus, by the beginning of the 1970s, it could be observed that the virtues of a policy centred systematically on producing for the world market in order to attain rapid growth and development had not been confirmed. Even if in a few countries the national product did grow substantially, that neither led to a real transformation of the structure of national economies, nor prevented poverty from growing and unemployment and under-employment from becoming established as scourges.

The most alarming consequence of this strategy based on the priority development of products in demand on the world market was at the level of the feeding of populations. Despite relatively high agricultural potential, food supplies in the region fell constantly, as figures supplied by the Economic Commission for Africa (ECA) show.

Thus, between the 1960s and the 1970s, imports of cereals rose by 38%, from 5.3 to 7.3 million tons. In spite of this, over the same period, the gross supply of cereals per capita fell from 160 kg. to 130 kg. Between 1961-63 and 1972, the volume index of food imports rose from 100 to 135, and the annual growth rate of imports over the period was 3.1%. By 1970, food imports accounted for 15% of the domestic production of the region, and 20% of food staples were imported.

The consequences of these imports on the external payments situation were naturally additional factors rendering the hopes of domestic accumulation for investment illusory. This policy, which is not neutral from the viewpoint of class interests, rested on the more thorough integration of the African economies into the capitalist world market and the intense exploitation of the peasantry in order to make this integration possible and consolidate it.

The world market itself has foundations which include an international division of labour and a system of exchanges based on inequality and directly related to this international division of labour. The unequal exchange of commodities between products supplied by dominated countries and by imperialist countries on the world market cannot be reduced to the deterioration of the terms of trade, which is simply its current visible and pressing aspect. Unequal exchange has more remote historical origins that must be sought in the relations that the capitalist economies established between themselves and the precapitalist economies of the countries that suffered a long era of domination and exploitation in various forms.

Whilst the vast resources extracted from these countries during the colonial period, and even before it, did not in every case start the process of accumulation in the capitalist economies, they did accelerate it. There ensued a tremendous development of the productive forces and levels of productivity in these economies, while the exploited countries subjected to this draining of resources saw their level of development drop and their socio-economic formations regress. Capitalism almost never sought totally to destroy the precapitalist modes of production that prevailed in these countries because, by maintaining them, it was better able to extract their substance.

Thus, during this long history, between the countries today called underdeveloped and the imperialist countries, a deep and unparalleled gap developed between levels of productive forces, levels of productivity and above all living standards of populations. Taking into account the very low level of the productive forces, the conditions of reproduction of the labour force in the exploited countries were reduced to an absolute minimum. Goods produced in conditions where reproduction of the labour force takes place at a very low level, command a very low price on the world market compared to the quantities of labour incorporated in those products. The result is a permanent and increased transfer of value from the exploited countries to the imperialist countries.

That is particularly true for agricultural products, the main commodities offered on the world market by most African countries because of the role assigned to them in the international division of labour established during the long history of domination. These agricultural products are bought on the world market at well below their value, that is, the quantities of social labour their production necessitated. Thus, the world market is the site where the super-exploitation of the African rural masses that produce these goods is realized. The phenomenon of unequal exchange is cumulative, and the deterioration of the terms of trade which is only one visible manifestation of it is growing year by year.

It is obvious that African political leaders who are endeavouring to involve the economies of their countries closely in this world market are middlemen who transmit the consequences wholly to the direct African producers. The fact is that prices at the level of African agricultural producers are generally fixed by government decisions. They take account of prices on the world market and the cuts taken by the African bourgeoisies who are actively engaged in international trade. And, from the circulation of products between African producers and the world market, the African ruling classes take a share of the surplus labour of the producers. Thus, integration into the world market turns out to be a very juicy business for these bourgeoisies. And if they denounce the deterioration of the terms of trade, it is simply in order to demand a larger share of the surplus labour extorted from the African producers, and thus consolidate their own economic base.

The consequences of this policy systematically directed towards integration into the world market have been disastrous, first and foremost for the African rural areas: massive and rapid impoverishment of the peasantry, accelerated degradation of soils, declining yields, and food shortages and chronic famine.

In fact, the almost exclusive encouragement given to cash crops brought a larger fraction of the peasant masses into the cash-crop economy. As a consequence, production of local foodstuffs fell and the incomes derived from cash crops were no longer sufficient for the people to keep themselves provided with foodstuffs because of the production costs incurred. Thus, by the end of the first post-independence decade, famine was widespread in a large part of the African rural areas.

Since the countryside feeds the towns, it goes without saying that the urban masses were not spared. The rural exodus and the flight of destitute and starving peasants towards the urban centres aggravated the situation.

This awful poverty and food disaster are the end product of the economic policy of euphoric extraversion of the first development decade.

We will be criticized for not giving enough importance to the particularly unfavourable climatic conditions of the late 1960s. Of course, these conditions affected the situation, but above all as an aggravating factor. The fact is that at the height of the drought in the early 1970s, which affected the Sahel particularly, the growth rates of export crops from the region were rising or marking time, whereas those of basic food crops were falling. Thus, between 1969-70 and 1973-74, percentage variations in the production of the seven[1] Sahel countries were −33% for millet and sorghum, −3% for groundnuts and +8% for cotton. The sharp decline in the production of millet and sorghum, the basis of the ordinary man's diet for a very large fraction of the population of the region, whereas production of cash crops was being maintained (as with groundnuts) or was increasing (as with cotton) can only be explained by the particular attention given to the development of the latter by the authorities (best lands in well-watered areas, hydro-agricultural installations, access to inputs, etc.).

The acute food crisis that resulted from the economic policy pursued during this period led to a greater dislocation of national socio-economic systems. Migrations, and above all urbanization, accelerated at a quite disproportionate rate since the supply of urban jobs was hardly rising. Massive unemployment was the result. The gradual depopulation of the countryside—as people moved elsewhere to seek some activity to enable them to survive or in the hope of benefiting from food aid—aggravated the decline of agricultural production which, in Africa, is based essentially on the massive use of labour power. For all these reasons, famine began to take on the appearance of a structural feature in most African countries.

How then could the threat this situation posed for the world system both economically and politically be dealt with?

In fact, vast starving masses of people are unlikely to produce in conformity with the objectives of the system. They cannot contribute to the accumulation of capital at the national and international level, nor assume the role assigned to them in the international division of labour. In addition, they constitute powerful political forces that can be mobilized with a view to the total overthrow of the whole system. This question came to preoccupy the monopolies, the multinationals, the international institutions and agencies that are heavily involved in the financing of the economic policies of underdeveloped countries, especially those in Africa, and which therefore have major decision-making capacities at the level of the orientation of these policies in the various countries.

The great crisis of the late 1960s thus precipitated a reappraisal that led to changes being proposed in the orientation of the economic policy advocated and implemented, especially in the essentially agricultural countries suffering from famine and deepening poverty. The proposals that were formulated and widely disseminated in numerous international and then national bodies were articulated on what was called the basic needs strategy, whose goal was to be the satisfaction of the basic needs of the great mass of the poor in the underdeveloped world. The main theme of this strategy, in so far as concerns the rural areas, was to consist in promoting the rapid development of smallholdings which provide a living to the vast mass of the poor in the rural areas of the underdeveloped world.

Thus, the so-called 'development of agricultural smallholdings' policy was tried out during the second post-independence decade in the framework of the basic needs strategy.

Notes

1. Gambia, Burkina Faso, Mali, Mauritania, Niger, Senegal and Chad.

5: The Second Post-independence Decade: The Food Crisis

The policy of developing smallholdings as well as the basic needs strategy of which it formed an essential feature were widely publicized by the World Bank, with all its vast financial power, which made itself their leading promoter. They raised hopes that were sometimes sincere, as well as many illusions. They left nobody indifferent, whether researchers or decision-makers. Given that Africa is in this regard the region in the world most implicated because it has the largest percentage of poor and rural-dwellers, we feel it necessary to look more closely at this policy as well as the strategy underlying it.

Smallholder Development Projects

According to its promoters, this policy consisted in giving financial support to projects aimed at developing production from smallholdings so that they could double their output by 1985.

While investment had hitherto been directed at big projects for developing tropical products intended for export, this was to be modified by paying greater attention to and giving an impetus to projects involving smallholdings, usually those producing food crops. Capital had therefore to be injected into traditional, mainly subsistence, agriculture with the aim of modernizing it and consequently of integrating it more into the world capitalist system. It is thus of considerable interest to examine some concrete cases of implementation of smallholder development projects, and their consequences on the social structures and conditions of existence of the peasant masses.

But before doing so, we need to clarify the notion of smallholdings by explaining their general features and the socio-economic conditions in which they develop.

General features of smallholdings

Given the great diversity of agrarian structures in Africa varying from one sub-region and country to another, the features which we are about to describe cannot be generalized for the whole continent. They do however apply to most cases in Africa south of the Sahara. We refer more particularly to the Sudan-Sahel zone for which we have more data and information.

Smallholdings play a very important role in African agriculture, both in terms of their number and in terms of the numbers of people they involve. Thus, in Mali, 75% of farms are under 5 ha.; in Kenya, in 1970, 770,000 smallholdings, ranging in size from about 0.5 to 10 ha. were recorded, while the number of large holdings was only 3,175. In Kenya again, in 1971, out of a total population of 11.7 million, 90% lived in the rural areas, 70% of them on smallholdings.

Smallholders are mainly involved in producing food crops, although in East Africa there is a relatively high number of smallholdings given over to producing tea or tobacco. But all that is a function of the needs of the family since smallholdings rely essentially on labour provided by the family to meet the family's needs.

Subsistence is thus the first aim and usually only surpluses are exchanged, or the product of plots devoted to commercial crops. Money and marketing thus play a minor role. For example, at Sédhiou, in Senegal, where a rice development project was launched to assist smallholdings, out of a rice production of 30,655 tons in 1976, only 5,000 tons, or one-sixth of the total, were commercialized. Actually, this example is a rather poor illustration of the limited role of commercialization since it relates to smallholdings that have already entered the process of modernization.

The labour is family labour and wage-earning is almost non-existent. But there are usually what may be called exchanges of labour that occur in the framework of mutual help groups. In turn, each member of the group invites all the others to get the work done faster. The one who does the inviting simply provides food and a few extras like cigarettes or soft drinks.

Techniques of production as well as farming methods are extremely simple and usually involve the use of agricultural implements and the application of the system of crop rotation with fallow, and sometimes use of animal manure.

The socio-economic contexts in which smallholdings develop relate to the main approaches to precolonial African societies which we outlined above. The social and ideological relationships inherited from those societies are not simply survivals. Some traits still remain almost intact in African villages. In general, land is still not the object of private appropriation, nor is it bought and sold. Access to land is assured in the framework of the village community. In a number of countries, however, a policy of privatizing land has been embarked on, which affects even traditional smallholdings (Kenya, for example).

If, in general, African tenure systems are marked by the co-existence of Western law and customary law, it is the customary law inherited from the old communal structure that mostly governs smallholdings. These customs determine the relations between men in dealing with land. They have rarely questioned the absence of private appropriation of land. The conditions in which production takes place (shifting agriculture) make a right of ownership unnecessary: it would be pointless. The absence of a right of ownership, however, in no way means the absence of any rights over land. The chief of the community (village or tribal chief) ensures access to land for members of the community. By clearing a given piece of land, these acquire rights over that land for their exclusive use. When it is exhausted, they can clear a new piece provided that they do not infringe on plots already brought under cultivation by others. Furthermore, transmission through inheritance which used to exist, still exists and goes through the maternal line (some coastal regions of the Ivory Coast, Ghana, etc.).

The assaults to which these tenure systems based on custom are sometimes subjected come from external interventions. This occurs when states decide to systematize the privatization of land. The explanation then given is to encourage peasants to invest in their land which they now own. Once this happens, land acquires a commercial value. This may also occur when an agricultural development project is started in a zone or region of smallholders.

Investments in terms of capital are non-existent or insignificant. Farm implements are produced by the peasants themselves or by the village craftsmen. Their extreme simplicity means that there is no separation between the producers and their means of production, just as also there is no separation between the producers and their products.

As for the general framework of the life of smallholders, if the main activity is agriculture in the narrow sense of the word, that is, the practice of cultivating crops, there exist in the traditional rural milieu many petty activities that vary from one ecological zone to another and according to the prevailing climatic conditions. For the most part, these activities involve crafts, herding, fishing, gathering and petty trade. The division of labour not being very developed, these activities are quite often carried on by the peasants themselves, either in the off season (crafts), or all year round (gathering, petty trade). Herding and fishing sometimes constitute the main activities of some small ethnic groups.

The social infrastructure is limited to a few schools and dispensaries sited in the main villages. Distance and inadequate means of communication and transport mean that the vast majority of peasant families scattered in countless small villages do not have access to the services offered by this infrastructure.

These are the principal features of smallholdings and their socio-economic contexts in a large number of African countries, especially south of the Sahara. It was necessary to enumerate them in order to be able to gauge the impact of the projects that are part of the basic needs strategy with the declared aim of modernizing smallholdings, and hence of transforming them.

Aims and Means of the Projects

While the declared ultimate aim is to achieve the growth of production, depending on the project, stress is put on various sectoral means. The projects usually concern food crops. Particularly in the Sudan-Sahel zone, they are aimed at increasing cereal production, sometimes in association with cash crops.

In other regions where peasant agriculture is more integrated into the money economy, efforts are directed at improving the production of some non-food crops, such as tea in Kenya or tobacco in Tanzania.

The means are varied: provision of credit to facilitate access to inputs; building of small hydro-agricultural dams; introduction of new agricultural methods and new varieties of seeds; preparation of new land through the eradication of illnesses like sleeping sickness; establishment of storage areas, etc. These means are used selectively, as required by each project; rarely are all used at once.

Some so-called integrated projects have as part of their aims to intervene not only in aspects touching directly on the development of production but also to promote social infrastructure like health services and education.

We shall see below how far these sets of aims were or were not achieved from the examination of a few concrete cases. Naturally, the concrete cases which we shall be able to examine are very limited in number, and our conclusions will be made with every reservation.

The number of people affected by projects relating to smallholdings varies a great deal, ranging from at least 5,000 individuals to some 500,000. It rarely reaches one million. And even then it must be stressed that these figures given by financing agencies are inflated to the extent that they are estimates relating to the population living in the project zones, but of which sizeable fractions do not effectively benefit from the real or supposed advantages. Constraints such as the conditions for the granting of credit or the inadequacy of irrigated areas, for example, eliminate many peasants who yet live in zones where projects are being implemented.

Structures for executing projects

The implementation of projects proceeds through the establishment of administrative and technical supervisory structures usually placed under the control of state bodies. In order to better understand the functioning of these structures and hence be able to assess their impact on peasant life, we shall take detailed examples from Senegal and Mali.

The administrative departments look after management problems, lay down the conditions on which land is allocated, especially if there is irrigation, lay down the conditions on which credit is granted, and ensure the provision of inputs as well as the collection and marketing of products.

The technical departments disseminate the technical aspects, teach new agricultural methods, make sure they are respected and particularly look after the time-table for their implementation.

This supervision to which the peasants are subjected sometimes takes the form of draconian constraints. For example, in the framework of the rice-growing project at Sédhiou in Senegal, and the agricultural development schemes in Mali (Operation Rice Ségou, Operation Rice Mopti, etc.) that we have been able to examine in some detail, the peasants had absolutely no say in anything to do with the management and marketing of their products. Thus, the Office National de Coopération et d'Assistance au Développement (ONCAD—National Development Corporation and Assistance Office) acted as intermediary between the peasants and the Banque Nationale de

Développement du Sénégal (BNDS—National Development Bank of Senegal).

The ONCAD was responsible for supplying machinery and credits to the Sédhiou rice-growing project. It also had a monopoly on the marketing of these products. Despite its ponderous administration and imposing bureaucracy which sometimes involved a lengthy delay in the actual delivery of machinery and the collection of products, seriously harming peasants' interests, the ONCAD granted itself an interest rate of 25% on credits, not to mention business charges. It is no loss that the ONCAD, so long lambasted, has finally been abolished.

The situation was almost the same in Mali except that instead of one body there were two bodies that shared the task of the ONCAD: the Office des Produits Agricoles du Mali (OPAM—Mali Agricultural Commodities Office) for marketing, and the Société de Crédit Agricole et d'Equipement Rural (SCAER—Agricultural Credit and Rural Machinery Company) for supplies.

Sometimes all the peasants in a village would be made collectively responsible for loans granted to individual peasants. Thus, as part of the Sédhiou rice-growing project in Senegal, credits were only made to peasants by the ONCAD if at least 85% of the old debts of the whole village were repaid.

In Mali, while the interest rate is not very high in the Ségou rice-growing scheme, there are compulsory levies in the form of obligatory payments, and compulsory sales which deprive the peasant of the bulk of his harvest since the levies can take over 50%.

In addition, the technical supervision is sometimes extremely restrictive. The least lack of respect for the new techniques being disseminated and the time-table for crops leads to the peasant's expulsion from the project zone and repossession of the plot that he was occupying. This supervision takes on the shape of a veritable police operation when it takes the form of the one on the rice-field at San in Mali. (Reproduced in the appendix is a request from the peasants there.)

The only rights left to the peasants in general are that they can have a say when there are, for example, hydro-agricultural works of concern to all to be carried out. They can propose these works, discuss how they are to be carried out and organize the necessary human investments. In addition, they can settle internal conflicts between peasants and discuss whether or not to accept new peasants on the laid out lands.

Impact of techniques introduced via projects on existing social structures

First, what are these techniques? They are generally superior to the existing agricultural implements and involve more demanding farming methods.

Some equipment can be kept by the peasants individually. These include: carts, ploughs, harrows, seeders, cultivators, etc. In the same way, inputs such as fertilizers, pesticides and selected seeds can be bought by the farmers as they need them.

Conversely, some heavy equipment is kept by the project administration which hires its services to the peasants. This includes: tractors, lorries, mechanical threshers, etc.

As for farming methods, they are applied under the supervision of the supervisors. Application of these methods is often made the first condition in allotting laid out plots.

These new methods impose some strictness in work conditions, use of certain selected seeds, strict respect for the agricultural calendar in the carrying out of the various tasks performed by peasants. The influence of the new techniques on peasants and village customs, traditions and habits appears in various guises.

By way of illustration we shall refer once again to cases in Senegal and Mali. At Sédhiou, in Senegal, more precisely in Casamance, the introduction of a certain variety of rice greatly contributed to reducing the sexual division of agricultural labour. Historically, the Mandinka people who settled in Casamance considered the valley zones as a place for women to work and the plateau zones as a place for men to work. Given that rice-growing was only practised in the valleys, the men never took any part in growing this crop, although the potential was enormous. They rather looked after the growing of groundnuts on the plateaux which brought in money as a cash crop. That was, of course, linked to the relations of domination between men and women in the framework of the family, the men looking after the crops that brought in money and ensured their power, and the women looking after the food crops that ensured the food supply.

Plateau rice had to be introduced to get the men interested in rice-growing. The pressure of the drought years meant that men and women would work together to lay out rice-growing lands. As soon as the men became interested in rice-growing through the introduction of plateau rice, the barrier of the sexual division of labour between ecological zones collapsed.

This disruption of an old tradition that hindered the development of the production of rice, the staple foodstuff of the region, can be considered a positive effect of the introduction of some new techniques. But the introduction of new techniques has other even more important consequences on family and social structures.

It is well known that, in the traditional rural world, the extended family embracing several households is quite a widespread type. This extended family has a large common field where every individual of working age from every household of the extended family works. The households may, however, have individual fields to which they devote a portion of their work time.

The acquisition of better implements rapidly leads to the spread of individual fields and the breaking away of households from the extended family. To the extent that possession of an ox plough and a cart makes it possible for a household to take care of the whole cycle of production without having recourse to members of the extended family, this household will tend to go and settle elsewhere.

Thus, the break-up of the extended family gets underway.

Examination of a Malian project, the Opération Arachide et Cultures Vivrières (OACV—Operation Groundnut and Food Crops) enabled us to observe that the development of individual fields was very clear in families having machinery as compared to those not having any.

Thus, in the pilot villages of the project, the following figures were noted: at Sirakémé, 88% of the fields are common fields and 12% individual fields in the unmechanized holdings, whereas in the mechanized holdings, only 37% are common fields as against 63% individual fields. In the village of Daban, 52% were common fields and 48% individual fields on the unmechanized holdings, whereas in the mechanized holdings, the common fields accounted for only 29% and individual fields 71%.

The development of techniques will thus tend to break up the extended family in favour of the family reduced to one household. Similarly, the village mutual aid associations will tend to disappear, as households manage more and more to carry out all the tasks of the production cycle without having to seek help from neighbours.

These phenomena occur even faster in projects where supervision and extension work impose strict respect for new farming methods. The new demands spelled out by the supervisors and the extension agents will tend to the development of independent work rather than work with the mutual aid associations in so far as plots are more clearly marked, and the constraints of methods and time more rigorous.

Thus, at the level of the family, the introduction of new techniques has major repercussions, but equally it has no less important ones on social divisions among the peasants. In fact, the plots with machinery save labour time and can offer their services to plot-holders who do not have any. In addition, the mechanized plots can increase their acreage and have recourse to outside labour that can easily be found among peasants who are short of cash.

Thus, in the framework of the OACV in Mali, we were able to observe that the number of working individuals at the level of a household rose with the acquisition of certain machinery.

We can already draw two very important conclusions concerning the introduction of new techniques in the framework of projects without prejudging their positive or negative impact. First, there is the retreat of the communal way of life and the tendency towards the break-up of the extended family and peasant solidarity. Then there is an acceleration of the phenomena associated with the spread of capitalism leading to the deepening of social divisions among the peasants.

Because of the uneven possibilities of having access to inputs as a function of the previous resources of different groups, this social cleavage will tend to deepen between peasants. Depending on the existing social legislation in the different countries, the better-off peasants can increase the size of their holdings in the framework of projects through loans, hiring or even sometimes purchases. The machinery that they hold enables them to exploit other land in areas outside the project. It enables them also to provide services for less well provided peasants.

These better-provided peasants thus have increasingly significant means of accumulation. Thus, a division of peasants into rich and poor is gradually coming about, as a result of unequal access to new techniques and unequal possibilities of exploiting these techniques to the full.

The projects and satisfaction of basic needs

Food needs: Quite often projects aimed at the development of smallholdings succeed in making significant increases in production, yield, productivity and even acreage.

For the Sédhiou, Senegal, rice project, realized production reached 2.8 times the targets by the second year, and 3.1 times the target by the third. For the Ségou, Mali, rice operation, while performances were less spectacular, they were nevertheless good. But while increases in production are generally achieved, in no way does this fact make it possible to conclude that the food requirements of the peasant producers are satisfied. For this there are three main reasons:

1) The share taken by non-producers (the state, suppliers of credit, etc.) can considerably reduce the amount remaining in the hands of the producers themselves.

2) The overall production figures for the project do not make it possible to estimate exactly the situation of individual peasant families: they do not always achieve the same results.

3) Even if each family satisfied its rice needs, for example, other food needs might not be satisfied and nutritional problems might arise.

The two rice production projects (Sédhiou in Senegal and Ségou in Mali) provide a good illustration of the scale of deductions. In the case of Senegal this deduction is effected through the fixing of a high interest rate for credit granted. This interest rate is 25%, payable at the end of each season. In Mali the interest rate is quite low (3%) but there are forced deductions in the form of fees and compulsory sales. These deductions may reach or even exceed half the peasant's production. This shows that no satisfactory conclusion can be drawn as to the situation of the peasants from the increase in production on the projects.

Our study of the OACV (in Mali) provides on the other hand an example of the fact that the global figures conceal disparities in situation between families. The OACV is a programme for developing groundnuts and food crops, mainly millet. The consumption norms deemed adequate by the Malian plan were 200 to 250 kg. of cereals per capita per annum.

The results of a survey conducted in four villages in the OACV put the level of consumption at 270 kg. per capita per annum for the villages as a whole. This might lead to the conclusion that the consumption norms were satisfactory. But an examination of the accounts of the holdings studied in the four villages revealed sometimes significant cereal deficits. Of 12 unmechanized holdings studied, seven had cereal deficits and purchases of cereals rose from 6,000 to 90,000 Malian Francs. Of 19 mechanized holdings, seven had cereal deficits and cereal purchases rose from 5,000 to 98,000 Malian Francs. This example proves that the global figures do not make it possible to get an accurate picture of the particular situations of individual families.

Finally, even if the cereal needs were satisfied for each family, other food deficiencies might exist because of inadequate consumption of some basic food staples: meat, milk, fruit and vegetables, etc.

In the final analysis the assessments and surveys that have been made in no way make it possible to say that basic food needs are satisfied by the implementation of the projects, either quantitatively or qualitatively. The only fair conclusion that can be drawn is that in the best cases the projects succeed to some extent in reducing food deficits. But even there, too, this reduction in the food deficit may occur at the national level and not at the level of the peasant producer, the latter producing more in order to feed other strata of society better, while his own food situation deteriorates (see the example of San in Mali, in the Appendix).

Educational needs: In the SDPs (smallholder development projects) the need for satisfying the peasants' educational needs is generally recognized. In reality, efforts are concentrated on training seen as technical supervision and popularizing new methods. Only limited means are devoted to schooling and literacy. For that there are no more than statements of intention and the appearances of action.

Technical supervision is naturally the main concern of the project promoters in so far as in their eyes it constitutes the main assurance of success. Thus no effort is spared to get new techniques accepted.

But for these same promoters the links between literacy and technical training are not always clearly perceived, or else the result of this literacy training appears so slow that it is not thought worth devoting substantial means to it.

In the projects which we have had the opportunity to examine, where literacy training is part of the programme, classes are given either in French or in national languages. The favourite topics are ones to do with production and productivity. But for whom?

Reports are rare.

But we do have one for the Ségou rice project (Mali). Between 1972 and 1976 results were 30% of initial targets. In the framework of functional literacy training the project envisaged that, by 1976, 600 centres would be opened, training 1,200 instructors for 70,000 students. Actual achievements were 190 centres, 359 instructors and 6,715 students.

But what seems even more serious than the insignificant place given to literacy training is that in none of the projects that we have examined have we found a programme providing schools for peasant children. To conclude, we can say that while references are made in the framework of the SDPs to educational needs, these needs are never really taken into account.

Health needs: The SDPs rarely concern themselves with peasant health problems. In the Ségou rice project (Mali) a number of actions are reported: between 1973 and 1976, 7,963 prenatal visits; 18,281 postnatal visits; 2,238 deliveries; 18,550 sick children cared for, and 74,579 malaria treatments. The total population of the project area was estimated at 150,000 in 1973. The results of these actions have thus been slight. But in most other projects these actions are not even mentioned and do not figure among their concerns.

It can only be deplored that the rural development projects do not make the building of dispensaries part of their programmes so as to ensure that the peasants have access to elementary preventive and curative treatment. Thus, the satisfaction of health needs does not figure among the goals of the SDPs.

Organizational needs: The requirements of organization pose for the peasants the whole problem of control over their own lives. This is thus a vital problem which for the peasants means control over the structures in which it is being attempted to integrate and supervise them. Only this real control can make it possible for them to transform and reorient development models conceived for them and outside them to serve their own true interests.

But this control is really possible only if the peasants constitute a force, are conscious of this force and capable of using it in order to reach their goals. In the last analysis the problem comes down to the mobilization of the peasants in well-structured and powerful organizational frameworks. It is therefore not to be expected that the promoters of the projects should help them to achieve this. The fact is that attempts to do any such thing would rapidly provoke conflict situations among the various parties interested in the project whose interests—to say the least—do not always converge.

It is not therefore surprising that in none of the projects studied did we observe the existence of this type of powerful organization. In the rice project at Sédhiou (Senegal), in order to facilitate contacts with the peasant world it was initially planned to set up village development committees. So 465 committees were created. They could have their say essentially when there were works of common interest to be carried out. They could suggest these works, discuss how to carry them out and organize the necessary human investments. In addition they concerned themselves with internal conflicts and whether or not to accept new peasants on newly irrigated land. Conversely, problems of management, credits, supplies, marketing etc., were outside their competence. The project did, however, express the desire that the peasants might in future be organized into co-operatives federated at the departmental level so as to take over for themselves all the administrative and supervisory operations which were assured by the administration of the project or the ONCAD.

The Sédhiou rice project is perhaps one of the cases where the desire has been declared to go as far as possible in the quest for solutions to the problems of peasant organization. In the Ségou rice project (Mali), the participation of peasants in decision-making is supposed to be assured by the fact that their representatives attend the campaign meetings and meetings of the management committee. Moreover, the project intends to carry on rural extension activities based on the community development centres and the functional literacy zones. To accomplish this, action committees have been set up at the village level and include local officials responsible for the project, and village and district delegates. The declared role of the committee is to seek solutions to the problems faced by the inhabitants. Rural extension groups have also been set up. Their members are elected by the general meeting of the villagers. Their role is said to be to ascertain needs, and to act in such a way as to ensure that the peasants participate actively in the life of the centres. In practice, it can be seen that control of the peasant world is the real concern.

These then are the various forms of organization proposed to the peasants in most smallholder development projects. In reality, the problems of peasant organization ought to go beyond the framework of the projects and be placed at the national level. The fact is that it is at the level of the central government that the most important decisions are taken affecting their basic interests, and this is done in their absence (fixing producer prices, credit terms, etc.). In order to be able to influence these decisions in their favour, the peasants would need to construct organizational structures that go beyond the village framework to bring together at the national level all the peasants as a whole on the basis of their common interests. Such a peasant movement would constitute a powerful force that could impose itself as interlocutor with a capacity to intervene, make decisions and react effectively to measures taken by the authorities. The process of creating such a movement is not a simple one in Africa, given that it tends to call into question the nature of the existing states and that it is likely to open up eventually the problem of the control of political power by the peasants.

Outside these independent mass movements which constitute frameworks for the struggle for the defence of their interests, all the structures established by the project administration and to which the peasants are sometimes obliged to belong (co-operatives, village committees, extension groups) are generally only means of increasing and extracting peasant surplus labour.

In the minds of its promoters the policy of developing smallholdings was intended to modernize and increase the production of smallholders. This increase has often been achieved. But the smallholders have drawn little or no benefit from it. The reasons for this are state levies for the benefit of other social strata or the exorbitant costs of modernization.

The fact is that, between 1975 and 1980, it was planned that the prices paid to Senegalese producers of peasant products would vary little or not at all, whereas the costs of inputs would rise appreciably. Thus, a ton of fertilizer would rise from 16,000 Francs to 25,000 Francs, an increase of almost 60%. The price of a UCF plough (a popular imported plough) would rise from 14,276 Francs in 1975 to 18,987 Francs in 1980, and a seeder from 16,392 Francs to 21,801 Francs.

Thus, in order to obtain the same implement or the same amount of fertilizer, the Senegalese or Malian peasant would have to provide a much larger quantity of his crop. It goes without saying that many of them would do without such modernization which in practice benefited only the state and the multinationals that produced these inputs.

The new techniques introduced have often destroyed the family social structures for mutual help and solidarity without always providing other balancing factors to compensate.

The social division of the peasants into rich and poor has been accentuated without either group seriously appreciating that they are together victims of exploitation by nation-states and big international capital. That sharply attenuates the class alliance and solidarity which ought to unite them since in the framework of smallholdings, even the rich peasants—we ought to have said well-off—have little chance of becoming petty capitalists. But the highest price of this modernization will have been to integrate backward traditional sectors more closely into the capitalist system with the serious consequence, as always, that a tiny minority does well out of it and the vast majority sees its living conditions deteriorate.

The basic needs strategy

The basic needs strategy is today well known. Studies made under the auspices of the World Bank, the ILO and many other agencies have widely publicized its content. We are not, therefore, going to fill a void at this level. In our section on the policy of developing agricultural smallholdings, we have attempted to raise a few practical problems that its implementation involves.

The strategy involves measures at the level of the rural areas and the urban areas, measures at the national and international level, economic and institutional measures.

We should like here simply to make a few comments on the problematic of the strategy as it relates to the rural areas, and its implications at the level of the whole capitalist system.

General observations

In the first place, the basis of the strategy is to seek to act on the situation of the poor with a view to improving it while at the same time preserving the underlying structures of society. But the poor do not live lives apart, but live within a society whose structures determine their conditions of existence. It is determinate relations of production that give rise to poverty and impoverishment and reproduce them at ever higher levels. Thus, without attacking these relations of production, one can only attack the effects and not the causes.

The strategy proceeds through the implementation of projects accompanied by a few proposals for reforms, which are usually quite vague. The projects are usually financed from abroad by big capital and are intended to relate to the sectors where the great disinherited masses live, that is, the vast traditional sector of the rural areas and the marginalized sectors of the urban areas. These sectors had previously been only indirectly or directly, but very loosely, integrated into the capitalist system. It is well known that wherever capital invests it tends to reproduce the relations of production that are typical of it and especially when the context is favourable, as it is in countries that are dominated and integrated into the system. It can thus be assumed that the implementation of these projects will lead to acceleration of the integration of these sectors into the economy governed by the rules of the system whose basic law is the quest for profit, and not help for the poor.

The financing of 'basic needs' projects does not undermine that of old type projects which incarnate the system of relations inherited from the colonial period. The extraverted general orientation of the economies thus remains intact, and it is to be feared that the food projects are simply means intended to sustain large-scale cash crop projects.

The capital invested is loans and as such these have to be repaid to those who set the original conditions. If the possibility of retrieving funds invested is not guaranteed, it is doubtful whether the investments would be made. But the sectors in which the impoverished masses live are sectors that are very little structured in terms either of institutions or infrastructure. This raises the likelihood of risks and leads the providers of funds to undertake detailed studies to select the lowest-risk situations. This inevitably leads to situations where cases of real impoverishment are left aside because they are not economically profitable.

If cases of failure multiply, it is to be feared that the strategy will be aborted. But the risks of failure are great given that the stress is put on the need for capital rather than on questioning the socio-economic structures of the countries concerned.

The reforms proposed to accompany the implementation of projects are very timid. But even with these minuscule reforms, how can there be any certainty that they will indeed be implemented and not taken over by the possessing classes. This leads one to wonder whether so-called projects for the poor would not profit the rich even more.

Development of the rural areas

It is in this regard that the literature on basic needs is most prolific, and it is indeed in this sector that the vast majority of the poorest people live, especially in Africa. We shall examine the overall goal proposed by the advocates of the strategy and then the means.

The goal: In his speech to the Board of Governors of the Bank, on 24 September 1973 in Nairobi, the President of the World Bank defined the goal thus: 'I suggest that the goal be to increase production on small farms so that by 1985 their output will be growing at the rate of 5% per year. If the goal is met, and smallholders maintain that momentum, they can double their annual output between 1985 and the end of the century.'

A preliminary remark, or rather a reminder, is that the estimates of the population affected by the basic needs strategy are inflated. In fact, these estimates relate to the population of the project area whereas a good proportion of this population does not participate in the project for various reasons (insufficient irrigated areas, credit inaccessible, etc.).

In so far as the farms affected by the project are concerned, is it logical to think that an annual 5% average growth rate can be attained between

1973 and 1985? There is no reason at all to think so. For many smallholders, their farms are generally not more than 1 ha. in size. Given the credit conditions—which we shall consider below—these farmers will find it difficult to have access to inputs. Not being in a position to use inputs making intensification possible, it will be impossible for them to sustain over a long period an annual growth rate of 5% of production on small farms whose fertility will rather be on a declining trend.

But even for the peasants who will effectively be able to reach 5% annual growth rate of production, how can one be sure that their standard of living will rise in consequence? In fact, it is known that most African states top up their budgets by creaming off the resources of the rural areas. The price systems, the tax systems, the credit system are all so many means that assure transfers of resources from the peasant producers to other social strata, not to mention the constant deterioration of the terms of trade between agricultural products and inputs. An annual production growth rate of 5% can perfectly well be accompanied by a deterioration of the conditions of existence of the peasants.

It is symptomatic to observe that the promoters of the strategy are only preoccupied with increasing production, without raising the problem of knowing who is to benefit from this increase.

The means: The means to be deployed are presented as follows:

Acceleration in the rate of land and tenancy reform.
Better access to credit.
Assured availability of water.
Expanded extension facilities backed by intensified agricultural research.
Greater access to public services.
And most critical of all: new forms of rural institutions and organizations that will give as much attention to promoting the inherent potential and productivity of the poor as is generally given to protecting the power of the privileged.[1]

But what does this catalogue of proposals, at first sight most appealing, cover? To find out, we shall look more closely at the details provided by other documents which analyse the proposals in depth.

Agrarian reform

In *Redistribution with Growth*, published under the auspices of the World Bank, it is stated: 'Where much potentially productive land in the hands of big landowners lies uncultivated, there may be widespread support for measures to turn it over to small farmers without a hint of an attack on rights to property.'

In *Employment, Growth and Basic Needs*, published under the auspices of the ILO, it is stated: 'Where large holdings are typically underutilised, agrarian reform may also be necessary to provide for the increase in food production required to meet basic needs.'

It will be noted that the proposals contained in these two publications, which did most to spread the word about the content of the strategy, conceive agrarian reform in the same way. First, the reform must relate to the unused lands of large estates. Next, it must be carried out with compensation for big landowners, as is spelled out in other passages. Finally, it must be carried out without challenging the right to property.

Given these conditions, it may be asked whether an agrarian reform subject to them is really applicable in practice and whether, in that case, there can legitimately be talk of true agrarian reform. In fact, in countries where the shortage of cultivable land is strongly felt, a redistribution of land which only related to the unused parts of large estates would be insignificant and could in no way ensure access to land to the mass of people needing it. On the other hand, getting round the measure would be easy for big landowners who wanted to, and they all want to. All they have to do is to act quickly to give the appearance of making use of unused land.

But even admitting that these lands can be recuperated, the 'fair compensation' mentioned poses problems. Who is going to pay the costs of compensation? If it is the poor landless peasants, it goes without saying that they lack the means, and what we risk seeing happening, as was the case at the time of the agrarian reform in Egypt in 1961, is that only rich peasants will be able to increase their holdings because only they have the means of meeting the expense of compensation. In this case, nothing will have changed in the living conditions of the poor and landless peasants.

If it is states that are responsible for compensating the big landowners, there is a strong likelihood that they will later pass on the costs incurred to the peasants. In this case, for the peasants who have secured plots of land, the running costs of farms will be such that they will continue to live in poverty.

In addition, is not the determination to respect existing property rights in contradiction to any real determination to carry out any agrarian reform? The response cannot be in doubt since wherever the problems of access to land are acute, it is the rich landowners who constitute the real obstacle. If their property rights are being respected, how could pieces of their property be taken away?

Those who advocate such reforms, however, think that 'almost all Third World countries have already adopted or at least promised to adopt legislation for this purpose.' They think that it is the implementation that is simply insufficient. This reveals clearly that an agrarian reform proclaimed by the dominant classes who control power, many of whom are big landowners, is only a charade with the political aim of fooling the people.

Advocates of the strategy declare their powerlessness to act on governments. They simply express the hope that the spectre of revolution will lead these governments to opt for the lesser evil. It is expressed in this way by the President of the Bank: 'But the real issue is not whether land reform is politically easy. The real issue is whether indefinite procrastination is politically prudent. An increasingly inequitable situation will pose a growing threat to political stability.'

Still with the aim of encouraging big landlords to make a few concessions to safeguard the system, other arguments are put forward in *Redistribution with Growth*. Thus, to counter the fear that an agrarian reform might lead to 'a sort of "landgrab" epidemic, accompanied by a good deal of dislocation and perhaps violence', the authors of this book explain that:

> More importantly, there is the general thesis that once the peasantry's immediate demands for land are met, it becomes a conservative force, a bulwark of the [new] status quo. Provided the process is controlled, therefore, a land reform in a setting where land ownership is initially highly concentrated can do much to alleviate the lot of the poor without tearing apart the fabric of society.

All that is very clearly set out and shows clearly that it is the preservation of the system that is the main concern.

But to say that the World Bank and other financial agencies do not have the means to put pressure on governments does not seem to accord with reality. In fact, the great majority of these states finance their economic policy from loans obtained from the Bank or other similar financing bodies. How could they resist conditions fixed by these agencies? Does not this confer on these agencies the most powerful means of pressure if they were truly moved by social goals?

It is difficult not to believe that there is complicity between the Bank and the powers that be in these countries. It therefore seems illusory to expect them to promote a true agrarian reform that would give land to the peasants. Such a reform can only come from the struggles of the peasants themselves, organized in powerful structures capable of mounting a comprehensive challenge to the famous status quo.

More credit facilities

Here, too, we need to go into some detail. In what spirit is the increase in credit facilities conceived? We shall turn to the proposals in the same sources. In the ILO's *Employment, Growth and Basic Needs*, it is stated: 'Water charges should generally reflect the true cost of providing irrigation facilities, and interest rates the real scarcity of capital. The prices of farm machinery should not in general be subsidized, and those of fertilizers only when this is clearly necessary to promote their use in the early stages of their introduction.'

The speech by the President of the World Bank points in the same direction; 'The fact is that concern over the usurious rates the farmer pays the money lender has led to unrealistically low rates for institutional credit.'

Thus facilities for access to credit are conceived according to the cold rules and mechanisms of capitalism. The approach consists in showing capitalists, notably those in backward linkages from agriculture, that smallholdings constitute outlets for them that have hitherto been unexploited, or insufficiently exploited. Then, in organizing credit systems open to smallholders with the greatest profit for the lenders.

It might be thought that if the small farmers find advantage in it for them, that might constitute progress as compared to a situation of stagnation or regression. It therefore becomes a matter of knowing whether the various imperatives mentioned are compatible, and whether smallholdings can be modernized through access to these credits.

Nothing is less certain. In fact in *The Design for Rural Development*,[2] another publication that appeared under the auspices of the Bank, Uma Lele reports the results of a number of surveys made in Kenya: 'The experience of the SRDP [Special Rural Development Program]'s Vihiga maize scheme emphasizes the importance of identifying the precise constraint to adoption before instituting a credit program.' 'Of the total of 600 farmers selected at random to participate in the Vihiga Maize Credit Program in 1971, only 54 qualified for credit under the Program's standards of creditworthiness. Adding another 22 eligible farmers from outside resulted in 76 potential borrowers. Only 63 farmers finally utilized loans.'

This example among many others shows that credit systems with their built-in demands are beyond the reach of peasants.

Our own experience of the study of a few projects has enabled us to observe very high interest rates (25% in Senegal). While on the agricultural projects in Mali the interest rate is relatively low, the state mops up peasant incomes through very high charges. In addition, the peasants are obliged to make compulsory sales of large quantities at very low prices to the state grain marketing office. When all is said and done, the Malian peasant working on irrigated land found himself short of food during the year (see the letter from the San rice-growers in the Appendix).

Thus, most of the time, credit systems are beyond the reach of the peasants. Those who have recourse to them often do so at the risk of food insecurity because of the debt repayment conditions.

In the last analysis, it can be said that the credit systems as they are proposed do not constitute a solution for modernizing smallholdings and improving the living conditions of peasants. On top of that there are the problems of dependence linked to the new techniques which these credits are intended to make it possible to acquire.

There will therefore be no choice but to turn to other alternatives.

Water supply

Stress continues to be put on large-scale irrigation works, even if it is felt that this needs to be complemented by smaller works. This shows that the lessons of previous experiences are not being adequately learned. Indeed in the same speech by the President of the Bank, it is stated: 'There are too many cases—in our experience and that of others—in which it has taken ten years or more after the dam was completed for the water actually to reach the farmers.'

It seems hard to believe that a dam can be completed and remain unusable for ten years. It is tempting to think that during that period of ten years when the peasant was receiving no water, the dam was being used for other purposes which were in fact the prime targets of the investors: for example to provide energy for mining industries controlled by foreign capital.

The experiences anyway show that a spread of small irrigation works, often built by the peasants themselves with very limited technical support, corresponds much better to the needs and means of smallholder agriculture.

These remarks allow us to say that the basic needs strategy is not a new strategy. It is only the continuation of old strategies of domination with the fruitless quest for escape valves to deal with the serious crisis affecting the whole capitalist system and particularly its weak link, the agricultural countries.

Given the particular features of the crisis, it is worth examining more closely the goals which this strategy sought to attain.

The basic needs strategy in the current capitalist crisis

From an examination of its implementation in the rural areas, it is possible to attempt to sketch out the particular role of the basic needs strategy in the current crisis of the capitalist system.

The capitalist system has entered a long-term crisis, and there is no sign anywhere to suggest that it will soon end. The manifestations of this crisis in the central capitalist countries are clearly visible: factory closures, record unemployment, wild inflation and revived protectionist tendencies. In the underdeveloped countries, they are the balance of payments crisis, the swelling of the debt with a high service element representing a large proportion of export income, and a significant reduction in import capacities. In short, the capitalist countries are no longer able to sell and the underdeveloped countries lack the means to buy. Thus, the crisis brings together all the conditions for a worsening of the tendency of the rate of profit to fall.

Indeed, the various capitalist groups will tend to increase the organic composition of their capital in order to be able to increase their productivity and make themselves competitive on the international market. The many-sided struggle between capitalist groups which results from this, and which is currently manifested in more or less covert protectionism between capitalist countries, is simply the struggle for the redivision of a world capitalist profit whose rate is tending to fall.

Added to that there is the excessively high external debt of the underdeveloped countries and those of the East, whence their growing difficulty in coming forward as buyers on the world market, which would have enabled production in the developed capitalist countries as well as international trade to pick up.

Usually, the developed capitalist countries were able to ride their crisis by stepping up the super-profits made on the back of the exploited countries, notably through the development of unequal exchange and the international division of labour, but it seems that this process is being blocked, the dominated countries no longer having the means to participate in exchange. As a result, the international division of labour on which unequal exchange rested needed to be modified. The system showed signs of needing new solutions to restore its balance.

The whole of the underdeveloped world could no longer be allowed to remain in the situation of non-industrialized countries. This was particularly important as a mass of technology was accumulating and there was no clear idea who to sell it to, and as the underdeveloped countries represented such a vast labour market that the factories would operate in these countries without excessive expenditure on wages. The size of the reserve army would ensure that wages remained very low.

Because of the need to get the system as a whole going again economically, it was thought wise to transfer some labour-intensive industries to the underdeveloped countries, and this is what was done. This shift was of course very unequal between underdeveloped countries, depending on their potential and the level of development they had already attained. But the hope for a new increase in the global mass of capitalist profit lay in this: a degree of industrialization of the underdeveloped countries based on the intensive exploitation of cheap labour in those countries. This exploitation would pose all the fewer problems because it would involve working classes which were unorganized or little organized, with limited political and trade union consciousness. They would therefore be insufficiently prepared to wage powerful struggles comparable to those of the working classes in the developed countries which were successful in imposing a certain wage level. What was more, the numerous totalitarian dictatorships in the underdeveloped countries acted as guarantors of the interests of capital.

But an important problem still remained: that of the continual decline in the food supply in a considerable number of underdeveloped countries. Wages are a function of the costs of reproduction of the labour force. The cost of food is the essential element in underdeveloped countries, and the food crisis, which led to the import of foodstuffs to feed the working classes in these countries, could not but lead to a substantial rise in wages there.

These foodstuffs necessarily cost more because they had been produced within the capitalist mode of production. In order then to minimize the costs of maintaining the labour force, it was necessary to reduce the cost of its physiological reproduction. Thus, it became urgent to increase food production rapidly and significantly. Local foodstuffs which cost far less would then be available for workers who would then be offered lower wages. Capital would thus be able to go on quietly making money.

It is, therefore, no accident that the growing interest in the food sector on the part of international financial institutions is occurring just when the trend towards transferring some industries to the underdeveloped countries is gathering strength. We believe the two phenomena are linked. It is, moreover, symptomatic that in the agricultural section of the basic needs strategy the emphasis is put only on increasing production, with no concern about knowing whether an increase in production can be accompanied by a decrease in the peasant producers' level of food consumption. This makes one very sceptical about the humanitarian argument used—'the fundamental case for development assistance is the moral one'[3]—and one might wonder whether improving the standard of living of small peasants is the real objective aimed at. It may be, however, that this aspect is not being wholly ignored. But in that case, does not the apparent objective hide other, political designs? We shall return to this point.

What seems certain is that the general behaviour of national states and financial and credit agencies towards small producers in the framework of projects to develop food agriculture does not make it possible to say that the satisfaction of the basic needs of small peasants is the central concern. The example of Zaire where, in 1971, the state obliged the big mining and transport companies to invest in food crops (cereals and legumes) with the aim of supplying the workers in the cities, strengthens us in that belief. It is quite certain that, in this state decision, the interests of the peasant producers are a minor consideration. Even if there is more investment in this sector, they will enjoy little or nothing at all of the results of this investment which will be creamed off and directed elsewhere.

But the economic objectives are not the sole concern of agricultural policy in the framework of the basic needs strategy. There are also political concerns. They are sometimes quite clearly expressed in the official literature on basic needs. The repeated warnings against the spectre of revolution to governments hesitating to make even the tiniest changes in the socio-economic situation of their country, the repeated statements that famine and poverty constitute threats to the stability of the system are all so many indications of the political objectives which can be summed up as the defence and consolidation of the system. How can this be achieved? The attempt consists first in trying to forestall the upheavals that foment under the scourge of famine and prevent the revolt of the starving which might result. Next, by seeking to promote among the small peasants themselves a small well-off stratum attached to private property and able to act as a support for the development of capitalism in traditional agriculture.

It is true that the situation of poverty and famine which is developing in the dominated countries in the periphery is becoming a nightmare for the leaders of the system concerned about its overall stability. It is obvious that any revolutionary upheavals that supervened in the countries of the periphery would be bound to have serious repercussions on the capitalist countries of the centre. So the strategists of the system put forward one proposal after another for action to defuse the bomb of poverty. But in doing so, they run up against the contradictions inherent in the system itself. For in fact the political objectives of defending the system as a whole are often at odds with the immediate economic interests and objectives of those responsible for putting these proposals into effect, i.e., the dominant classes of the national states and the providers of funds. This contradiction takes the following form: how to fight against poverty and reduce suffering, while at the same time ensuring substantial profits for foreign investors and guaranteeing significant transfers to the benefit of these dominant classes and their bureaucracy, and all that through the implementation of development programmes based on the super-exploitation of these same poor people?

This is a fundamental contradiction. And despite the questions begged and the declarations of good intentions, the structures of the system end up imposing their laws, which have nothing to do with the 'moral' argument. The problem therefore remains one of knowing how to guarantee the stability of the system while reducing the threat to its periphery.

For the time being, since it is not possible to improve the lot of the poor, the attempt is being made rather to silence them by establishing or maintaining political dictatorships which at best manipulate the poor masses, and at worst purely and simply massacre them.

As for the other option which would consist in wanting to promote, within the mass of peasants, a well-off fraction capable of prospering within the system as it currently functions so as to be able later to defend it, that remains within the realm of the possible. We have, in fact, observed that a small minority of peasants who are rather better off before the implementation of projects sometimes succeeds in deriving substantial advantages from the execution of these projects. They can acquire means of production (a plough, carts, oxen, a seeder, etc.). They can enlarge their farms, or hire out services to peasants in difficulties. In this way they accumulate, which enables them gradually to separate themselves from the mass of the peasants through their income level and their conditions of production. They are naturally attached to private ownership of the means of production and there is no doubt at all that they will fight fiercely to preserve it. They constitute links in the system, and it is hoped that they will acquire sufficient political power to play the role of leading the great mass of peasants and involving them in the defence of the system.

But it seems more likely that things will turn out otherwise: a mass of peasants getting poorer by the year will surely end up openly fighting against a minority that gets richer by the year with the support of the leaders of the system. Only the ensuing class struggle will open up the hope of a radical reordering of the system.

To conclude, it could be said that the capitalist system has experienced regular periods of crisis of varying importance, followed by periods of a return to a certain equilibrium under the impact of measures that are often simply conjunctural. This system is currently experiencing a prolonged crisis which is manifested in various forms and is eating it away at various levels.

The basic needs strategy is part—even though in a very secondary manner—of the panoply of ways and means being tried out to rediscover this equilibrium which is never anything but an unstable equilibrium. These few comments from an examination of its agricultural section will perhaps have made it possible to bring out, to some extent, the insurmountable contradictions with which the system is confronted.

Notes

1. Speech by the President of the World Bank, Nairobi, 1973.
2. Uma Lele, *The Design for Rural Development* (Washington D.C., Johns Hopkins University Press for the World Bank, 1975).
3. Speech by the President of the World Bank, Nairobi, 1973.

6: Forms of Control

The forms of control of the system of production by capital can be highly varied. They can, however, be put into two broad groups:

1. Forms of control of production by means of controls of the price and distribution systems leading to particular types of subordination of the peasants, that vary from case to case.
2. Capital inserts itself directly into production and openly exploits a wage labour force. This is typical agricultural capitalism. Capitalist relations of production are present here, whereas they were not in the first case.

Between the first case in which the peasant still retains some degree of autonomy, although being subject to some extent to capital, and the second case in which the peasant is nothing but an agricultural labourer who sells his labour power, produces surplus value for the agricultural capitalist and collects his wages, there are other situations in which the peasant occupies intermediate positions, sometimes closer to those of the autonomous peasants, sometimes to those of the agricultural labourer.

The first case, in which capital limits its activities to an indirect control of production, is the commonest one in sub-Saharan Africa. The peasant has his own land and his rudimentary instruments of production. He produces for his own consumption and partly for the market. He may also produce essentially for the market, i.e., by devoting himself to cash crops and hoping to be able to buy foodstuffs with the income from these crops. The provision of inputs and the collection of products are carried out by other structures or other social strata. The colonial administration and trading companies that performed this function during the colonial period were replaced after independence by a new national administration and structures. Local traders occupied, and sometimes still continue to occupy, a major position. Similarly, to a lesser extent and more recently, transnational corporations are directly present in some states. The activities over which the peasant has no control at all cover upstream provision of credit for inputs and sometimes technical back-up. Downstream, they include the collection of products and their transport to local processing plants or their sale abroad.

These national or foreign agencies, as well as the social strata that control these activities, are also those that dominate the fixing of prices, in both major risks. These risks include the uncertainty of the weather, costly clearing work, high cost of inputs, etc. These are all factors which can raise the costs of production and reduce the possibilities of profit. Capital thus has an interest in leaving it to the peasants to work the land directly and so bear the risks that involves.

We are here dealing with the following situation: peasants, without being dispossessed of the land and while keeping their means of production, lose control of their production to the capitalist sphere which dominates the distribution system. How then can the status of these peasants be described? Are they proletarians as some have suggested? Can one say that these peasants are producers of surplus value? This question relates to theoretical considerations which have been the subject of sometimes impassioned debates. The school of thought that is opposed to the classical positions of Marxism, or the supposedly classical positions, considers that the peasant so exploited is simply a *de facto* proletarian producing surplus value. Behind the appearances, behind the sale of products, lies concealed the sale of labour power. According to this school of thought, it is necessary to go beyond the immediate appearances and legal relations—the peasant owning the land and those rudimentary means of production—and consider the economic relationships at the level of the system as a whole in which the peasant is closely integrated. Then, it would be realized that he simply sells his labour power to capital, a sale that is dissimulated in the form of a commodity sale.

In short, he is seen as a proletarian. What should be made of the thesis thus advanced by this school of thought? Is not the small peasant who has his own land and his instruments of production, but is exploited in some way by capital, simply a proletarian, and hence a producer of surplus value? We shall try and make a few observations arising from the questions raised by this thesis.

In our opinion, the danger of this thesis is that it reduces to a single form the manner in which capital exploits labour. This single form is seen as consisting in the sale of labour power and the production of surplus value. Anyone exploited by capital would then be a proletarian.

But the notion of production of surplus value attached to the appearance of a clearly determinate class, the proletariat, is a very precise category in Marxism. It presupposes, on the one hand, the free worker owning only his labour power and, on the other, the capitalist owning the means of production, being able to buy this labour power and use it for more than the time socially necessary for its reproduction. The difference between the labour time necessary for the reproduction of labour power and the time the capitalist uses this power constitutes surplus value. Once he buys it, the capitalist has the labour power and uses it for the production of commodities.

For selling his labour power the worker receives a wage which corresponds to the means of reproduction of the labour force given the socio-economic characteristics of the country where the sale takes place. The worker thus called a proletarian is a producer who is separated from the means of production and separated from the product. The labour that he performs is generally part of more collective labour.

The type of relationship that exists between the small peasant and the capitalist does not seem to us to be of the same nature as the type of relationship that links the proletarian to the capitalist.

Although both workers, small peasants and proletarians, are exploited and subordinated to capital, the degree of subordination is not the same, just as the dispossession of the labour process is not the same.

The small peasant who is exploited by means of the exchanges of commodities that occur between him and capital—a merchant capital—can, when he feels that the extortion of surplus value is being intensified, abandon, at least in part, the production of commodities sought by the capitalist market. He can thus reduce considerably, or even totally, the labour time that he devotes to the production of commodities sought by that market in favour of, for example, subsistence production. Similarly, he can give up the inputs offered by the capitalist market once he realizes that these inputs cost him more than they give him.

Occurrences like these, which can be observed frequently in the Sudan and the Sahel in Africa in particular, where peasants can be seen giving up inputs and abandoning cash crops to take up food crops, prove that they retain a considerable freedom of manoeuvre. The small peasant has this possibility of regulating his commitment to the system by reducing or increasing his degree of integration, but the proletarian does not.

Of course, even the peasant who produces only foodstuffs finds himself exploited by capital in the sense that he is obliged to market at least part of his production because he lives in a money economy and he needs cash to meet certain obligations. In doing so, he sells his products at very low prices to non-foodstuff producers who in general work for the capitalist market and are poorly paid, in particular because of the low prices of agricultural products. Thus, the non-foodstuff producers find themselves exploited and through them, the small peasants are too.

The peasant can even find himself in a situation where he is distinctly more exploited than the proletarian by the fact that these products are bought for far less than their value than is the labour power of the proletarian. Thus, without being a producer of surplus value, the peasant can have more of his surplus labour extorted from him. That in no way alters the difference in kind that exists between surplus value and this form of extortion of surplus labour. Some further observations can be added with regard to the conditions under which they work and their ideological and political consequences.

In capitalist production that generates surplus-value the labour process is collective, whereas it is individual or family-based in small peasant production.

Furthermore, insisting that the small peasant is tightly integrated into the capitalist system may create the impression that he now plays a completed and unalterable role in the general functioning of the system, whereas the status of small peasants can evolve in the direction of kulakization or in the direction of the deterioration of the conditions of existence leading to rural exodus and proletarianization. Finally, the mere ownership of the land and the means of production, however rudimentary they may be, leads to the peasant behaving and having a set of reflexes that belong to a different ideology from that of the proletarian: an individualist ideology underpinned by the feeling of having the possibility of becoming a kulak.

Ultimately, we feel that so long as capital does not get involved directly in the process of production, the peasant producer suffers a very severe extortion of surplus labour, but it is different in kind from the production of surplus value. The extortion of surplus labour occurs in the form of a relation of distribution, whereas in the creation of surplus value a relation of production is involved. This possibility of getting out of the structures of the system makes us think that the subordination of the peasants is rather informal. The still informal form of subordination that the modes of production to which the small peasants belong are subjected to usually leads to their regression.

In the case that we have just considered, we stressed particularly the situation of the small peasant integrated into the system, but still retaining a degree of autonomy which is manifested in the possibility of disengaging, at least partly, from the structures that make this integration possible. The case is widespread in the Sahel (Mali, Senegal, Niger, and Burkina Faso especially) where groundnut- and cotton-growing, the main base of the trading economy in this region, can suffer marked falls when the conditions of existence of the peasants engaged in it seriously deteriorate.

Compared to this situation there are others where the form of integration is much deeper, with capital getting involved directly in production. This is the system of small peasants, or kulaks, mainly in the plantation economies, state-irrigated lands worked by peasants who are like agricultural labourers, and finally systems of production maintained by multinationals. We think it can be said that in these cases we are really dealing with a capitalist agriculture, even though nuances need to be introduced. For, between the wealthy cocoa planter in the Ivory Coast and a transnational producing out-of-season fruits and vegetables, there is a vast difference. Generally speaking, in these cases the farms function as capitalist enterprises. Communal appropriation of the land is disappearing and giving way to private ownership.

The wealthy planter, the state owner or the foreign company appropriate large tracts of land on which a large wage labour force works. Production is almost wholly destined for the market. The means of production, which are the very latest, require considerable investment.

Let us examine these three cases in which capital intervenes in the process of production, which involves links between the peasants and these various capitalists: the wealthy peasant who is generally a planter, the state owner and the transnationals.

The wealthy peasant or planter, who is often the descendant of old traditional chiefs, has a wage-earning labour force made up of agricultural labourers. Around him lives a mass of often poor, independent peasants. This wealthy peasant or planter, who we shall call a kulak, enjoys great influence among the peasants. This influence shows itself in his tendency to occupy leading roles in peasant associations or co-operatives. With a capital stock in machinery and money, he exploits a wage-earning labour force. He hires out his services and can make loans at often usurious rates to the independent peasants who are to some extent dependent on him. He naturally has the reflexes and behaviour of a landlord.

For big capital and the neocolonial state, he represents a natural ally and the motor for the integration of the peasantry into the system. Since he finds himself in a peculiar situation compared to the rest of the peasantry, the vast majority of which remain poor, there develop between him and the rest contradictions of a complex nature. The rest of the peasantry has both a tendency to be hostile to him because of the economic power he possesses that enables him to exploit the rest of the peasants, and to seek his favours because of the political clout he possesses at the level of the country's leadership. But the tendency will be rather for the antagonism to grow.

In fact, the process of the development of capitalism will lead to a continued impoverishment of the mass of peasants, and this impoverishment will go hand in hand with the enrichment of the kulaks. There will be cumulative indebtedness on the one side, and increased accumulation on the other. Eventually, the kulak will tend to enlarge his holdings by buying up those of ruined peasants, who will be transformed into agricultural labourers or migrate to the towns.

The kulak is situated in the capitalist sphere. From his agricultural labourers he draws only less surplus value than the absolute surplus value in relatively little developed conditions of capitalist relations of production. The agricultural labourers here are in a situation of formal subordination.

From the mass of peasants he extracts surplus labour in the framework of relations of distribution. There again we are dealing with the type of informal subordination previously examined.

This situation, which brings together the conditions for primitive accumulation, is common in East Africa and southern Africa where the system of land registration leading to its private appropriation is well advanced. Variants are also to be found in the plantation areas of Central and West Africa.

In the examples of development of a capitalist agriculture, we have also mentioned the case of lands irrigated by the state on which peasants work in conditions which are sometimes such that they are effectively agricultural labourers. This refers to certain forms of intervention by the authorities in the rural areas, where the state practically plays the role of owner. Thus, the state, for example, lays out irrigated lands on which it puts peasants to work according to conditions that it alone determines. These interventions, whose aim is always said to be the improvement of the peasants' living conditions, thus become means of extracting surplus labour.

Usually, the state allocates the peasants plots of irrigated land which vary in size according to the size of families and to what means of production are held. On these lands, the peasants are required to grow specified crops.

A quite dense administrative and technical network surrounds the various stages of the whole process of production, which is totally beyond any control by the producers. The administrative structures provide management while the technical structures determine working conditions. Management includes supply of inputs, allocation of credits, collection of products and securing repayment of debts. Prices and interest rates are wholly the responsibility of the administration which is a state structure. The technical staff dictates and controls the application of technical methods: timing and type of soil preparation, timing of sowing and weeding, use of inputs, etc. The peasant is subject to a strict agricultural calendar and cropping methods. If he fails to respect these conditions his plot will be taken away from him. The peasant who has to live with this situation is not the owner of the land. His plot is leased to him, usually in exchange for payment of a fee. He does not own the inputs, which he is often obliged to use. The credit conditions that enable him to acquire these inputs border on the usurious. He is not free to choose his own production methods, but must accept those imposed on him. Nor does he own the product since he only receives a fraction of it after various deductions have been made for fees, credit repayment, etc.

Thus, in many respects, this peasant finds himself in the situation of a proletarian who simply sells his labour power and receives his wage in kind. What is more, he is obliged to sell all or part of the portion of the product that he does get at a price far below its value.

If the peasant here does not become a full-fledged proletarian with a wage set in advance, it is again to make him bear the risks of a bad harvest. In this system, in which the state receives a portion of peasant surplus labour, the rest going to central capitalism through the operation of unequal exchange, the peasant is super-exploited. The payment for his labour is in no way related to the conditions of reproduction of his labour power. We are dealing here with an agrarian state capitalism which reduces the peasantry to a supplier of cheap manpower. The peasant simply sells his labour power, for which he is paid in kind. He is an agricultural worker inserted in little developed capitalist relations of production. Here, too, he suffers a formal subordination.

Finally, in the examples of development of a capitalist agriculture, there are the cases of interventions by transnationals and agribusiness. The transnationals which are involved in agriculture attempt to dominate all aspects of supplying inputs, purchasing products and processing them, seeking out and controlling markets, fixing and manipulating prices.

However, intervention by transnationals in African agriculture is not yet on a massive scale. In many countries, they are active in small areas or are still at the experimental stage. Compared to the colonial companies of the old days, their interventions have a number of peculiar features.

First, they do not hesitate to make or call on large investments as has been shown by BUD (Bud Antler Inc. a multinational, based in California, dealing in African business) in Senegal. BUD brought in highly qualified technicians from Europe and America to set up an irrigation system based on state of the art technology: drip irrigation with the water brought by pipeline in a perforated tube over long distances.

directions—both for the goods supplied by the peasants and the goods supplied to the peasants, the latter usually not being subject to price-fixing. Thus, they are subordinated to capital which extracts from them a significant amount of surplus labour through control of all the distribution circuits.

Dominating the system of price-fixing, the capitalist sector tends not only to keep the prices paid to producers as low as possible, but at the same time substantially to raise the prices paid by the same producers to purchase inputs.

The explanations given to justify this double extortion to which peasants are subjected are always the same: to cover transport and management expenses and international competition. But the reality is that on the basis of this intense exploitation, capital flourishes and the dominant classes operate a major transfer of peasant resources to their own benefit. Some examples from the Sahel zone will illustrate these extractions of resources to which peasants are continually subjected. Between 1965 and 1974, in ten years, the producer price of groundnuts rose by only 16% in Chad, 13% in Senegal and not at all in Burkina Faso, whereas over the same period the price of groundnuts on the world market had risen by 108%.

For cotton, over the same period, producer prices rose by 23% in Chad, 6% in Senegal, 13% in Niger and 3% in Upper Volta, whereas the price of cotton on the world market had risen by 118%.

Between 1967 and 1973, the Gambia Marketing Board, the state agricultural produce marketing board, moved from a loss of 1.5 million Dalasis to a profit of 9.7 million Dalasis. But producers' income rose by only 39%.

Between 1967-68 and 1972-73, the value of Senegal's groundnut production fell from 22.1 to 20.5 billion Francs CFA. The producers' share fell from 66% to 43%, whereas the income accruing to state bodies rose from 7% to 41%.

These are examples taken from public authorities which are not the only ones involved in the process of extracting surplus labour from the peasants through the price and distribution systems.

Thus, without being dispossessed of his land and his instruments of production, the peasant finds himself in a situation where he is dispossessed of his production itself. The logical consequence is that the peasant quite often simply abandons the inputs and even cash crops and devotes himself almost entirely to growing foodstuffs. Instead of the intensification encouraged by the suppliers of inputs, the peasants prefer an extensive agriculture which, despite its ecological consequences, guarantees them less dependence and fewer debts. This altogether rational behaviour—many researchers argue the reverse—leads to the rapid impoverishment of soils. Caught in the vice of excessive indebtedness and low productivity leading to the same shrinking of incomes, the peasant often finds himself forced to leave the land and migrate.

But capitalism is not interested here in getting hold of the land, just as it did not seek to alter the tenure system so as to dispossess the peasant of the land. This is because, in general, in these cases, land is relatively plentiful—albeit often impoverished land—which means that land has no commercial value, or only a very low one. Then, direct working of the land here involves

They are also less concerned with actually owning the land, knowing that whatever the type of ownership, they can succeed in controlling the whole system of production and distribution. In addition, non-ownership of the land reduces fiscal problems and the risk of nationalization.

Again, as in the past, they are interested in export products, sometimes even in very specialized products such as out-of-season vegetables sought by consumers in the industrialized countries. But they are not only interested in this. They sometimes attempt to organize local production to supply African markets which they thus control. This is the case, for example, with ranching which is developing in some West African countries with large herds of livestock.

Finally, the contracts that bind the transnationals to African states are built on unequal relations of force. Consequently, they can impose conditions that seriously affect national resources: occupation of the best lands, relentless exploitation of African manpower, taking over of some of their setting-up costs by local public finance.

In the last analysis the transnationals and agribusiness seek to combine for their own greater profit, technology—sometimes state-of-the-art technology—with the cheap labour force and the natural advantages offered by African agriculture, while also securing the political guarantee of national states. We shall use the example of SODEPALM, in the Ivory Coast, an agro-industrial group built up around BLOHORN, a shipowner, and owner of an oil-palm company in the Ivory Coast. BLOHORN owned palm plantations worked by agricultural labourers. At the same time, it controlled the production of village planters. It processed the products into oil; with the development of the group, made possible by participation by the state and the injection of capital from external sources of finance, the activities extended to oil products and to a whole new range of processing operations: table oil, soap-making, margarine, detergents, etc. This is then an agro-industrial complex with overwhelming economic and financial power compared to the average village planters whose production it collects. These planters exploited 26,200 ha. employing 6,000 peasant workers.

BLOHORN and SODEPALM directly exploited 42,000 ha. worked by agricultural labourers. The domination of agro-industry by the BLOHORN group is such that it makes any possibility of independence for the village planters illusory. In these conditions, peasant agriculture exists only in name. The village planters have no possibility of getting out of the system that grips them. Decision-making powers belong to a power beyond any control by wage-earners and planters. We are dealing with a vast industry which determines salary conditions and income structures in the whole system of production and distribution.

Thus, in the agro-industry controlled by BLOHORN in the framework of SODEPALM the various forms of subordination of labour to capital are present. Also present are relations of distribution, little developed relations of production and fully developed capitalist relations of production. The small planters who sell their products to SODEPALM are in a situation of informal subordination, which is very close to the formal subordination of agricultural labourers in the sense that their capacity to disengage themselves from the system like the small peasants in the Sudan and the Sahel is greatly reduced. Their degree of autonomy is weaker. Nevertheless, it can be said that they are still integrated into relations of distribution with capital, with stronger domination by capital. The agricultural labourers who work in the oil-palm plantations of SODEPALM maintain little developed capitalist relations of production with capitalism, and suffer formal subordination.

Finally, the manpower that works in the processing industries produces every form of surplus value (absolute or relative) given advanced techniques. It can be said that they are integrated into fully developed capitalist relations of production. Their subordination to capital is complete and entire. It is a real subordination.

To conclude: from the colonial period to the present day, African agriculture has undergone and is still undergoing a great variety of changes, changes which have led to various forms of domination by the capitalist mode of production over the pre-capitalist modes of production that previously existed. This domination led to a retreat of precapitalist modes of production, without however the dislocation being total; such dislocation would not be in the interest of capitalism. In fact, it is through maintaining precapitalist modes of production that capitalism can ensure cheap agricultural production perpetuating both the conditions of unequal exchange and those of low pay for labourers working for capital.

During this long process of transforming agriculture, the peasantry was integrated into the system in various forms, ranging from informal subordination—the peasant, although exploited, retains some freedom of manoeuvre *vis-à-vis* the structures of the system—to real subordination in which the peasant leaves the land for the factories of agro-industry. He then enters into fully developed capitalist relations of production. He becomes a worker producing both absolute and relative surplus value.

In terms of these two statuses, that of formal subordination corresponds to that of the agricultural labourer in which the worker, although producing surplus value, finds himself integrated into as yet little developed capitalist relations of production, with a limited level of technology that does not allow surplus value to be extracted as fully as possible.

The intense extraction of surplus labour to which the peasants are thus subjected, whatever their forms, lead to growing poverty and a profound crisis of African agriculture, a crisis that is reflected at all levels of the economic and social life of African countries.

The problem that remains posed is whether an alternative exists, and what the conditions for its existence would be.

7: The Alternative and its Prerequisites

In the previous chapters we attempted to demonstrate, first through figures, what is today beyond any doubt: that African agriculture is in the midst of a serious crisis. We have demonstrated that the origins of this crisis go back to the earliest contacts between European capitalist economies and African precolonial, precapitalist economies. We recalled that the colonial period led to a disintegration of peasant economies, and stressed the fact that the first post-independence decade, through the policy of increased extraversion conducted by the newly independent states, precipitated the crisis by systematically orienting the national economies to the world market. We observed that the solutions to the crisis put forward during the second post-independence decade were useless and illusory. Finally, we tried to show that, throughout this period of domination which took various forms depending on the period, the peasantry, the basic productive force, bore most of the burden; and that it suffered and continues to suffer the exploitation of capital in various forms of subordination.

It is by looking at these facts that we can reflect on the possible alternative and the conditions for this alternative. It is obvious that solutions cannot be sought only at the level of the rural areas, nor even only at the economic level. The crisis which we are living through is a political, economic and social crisis which calls into question the whole of the existing structures in African states, and the types of relations linking them to the world capitalist system.

By posing the problems thus, we are bound to place ourselves in the context of the class struggle at the national and international level which determines the policies thus followed. And without being thoroughly acquainted with the interests that bind these classes together or the struggles that their opposition arouses, it is impossible to pose correctly the problems of the alternative. These call for a clear view of the attitude of each class to the crisis which does not in reality affect everyone equally. Thus, in order to reflect lucidly and objectively about the alternative, it is essential to pose a number of problems, the terms of which situate the crisis and alone make it possible to envisage possible solutions.

In order to grasp these crucial problems, which necessarily bring into play class interests in contemporary African societies, it is necessary first of all to be able to appreciate the class composition of these societies. Without being fully informed of the social processes at work, one cannot know why this or that policy is carried out and persisted in, despite these negative consequences.

This, therefore, implies having a reasonably clear idea of the class nature of the existing political rulers who have a vested interest in the harmful policies hitherto pursued. Next, it is necessary to think about the popular class alliances that might lay claim to political power and carry out a different development policy. Finally, it is necessary to look at what other development policy the popular alliance could carry out. This can only be sketched out, given the great variety of situations in African states, the variety being linked to a whole series of economic, ethnological, sociological and other problems.

Hitherto, so far as we are aware, there has rarely been an analysis of the alternative in these terms, and that is so not because the need for such an analysis is not felt, but because its extension to Africa beyond the framework of a single country, given the numerous differentiating factors mentioned, makes it difficult to carry out. It is important to undertake such a study, however, even though it may suffer from inadequacies or weaknesses.

Evolution of social stratification in the African states which achieved independence in the 1960s

This stratification itself is the product of the impact of capitalism on precolonial and colonial traditional societies. The most striking impact has been, if not to engender, then at least to exacerbate, the exploitation of man by man within African societies. Thus, at the time of independence, a fraction of the society occupied economically dominant positions, made up for the most part of those who had thrown in their lot with the colonial power. It got a share of the surplus labour extorted from the toiling masses and its political ambitions were strengthened. This fraction included essentially bourgeois strata, bureaucratic strata, and 'feudal' forces, which were usually agrarian and religious.

The bourgeois strata included essentially big traders acting as middlemen between big capital and the masses, and big planters in varying numbers depending on the geographical zone. The African industrial bourgeoisie was still practically non-existent (except for Egypt and South Africa?). The bureaucratic strata included the former African representatives in French metropolitan institutions (Assembly, Senate and even government), officials in the colonial administration and army, and paramilitary forces' personnel. The 'feudal' forces included essentially the religious forces that were sometimes very powerful in the rural areas, both economically and politically.

Alliances of interests generally brought these strata together; they dominated the political sphere at the time of independence even though latent contradictions existed, notably between the petty bourgeoisie of officials without an economic base and the bourgeois strata already possessed of solid bases.

At the other end of society, there was the other fraction which included the mass of small peasants, the small minority of the working class (except in a few rare countries which have marked peculiar features where it was already numerically important: Egypt, South Africa, but their independence does not date from the 1960s). Alongside these two groups of classes, there was the stratum of low-level employees, artisans, etc., the great majority of whom had no well-defined stable job.

The first fraction, dominant on the eve of independence, was often successful in securing the political support of the second to negotiate independence to its advantage. But this was not always without violent conflict.

In fact, there were sometimes sharp struggles within the first fraction despite alliances, which were sometimes only temporary, to determine which stratum would inherit from the colonial power. And these struggles were reflected at the level of the masses with the same intensity, more through manipulation than through class interest.

Thus, in many countries the petty bourgeoisie of officials which was to constitute the backbone of the bureaucratic bourgeoisie confronted bourgeois strata that had an economic base and were more closely linked to the colonial power.

Where this petty bourgeoisie succeeded in taking power it was in general to install a state-run economic system, the means of providing itself with its own economic base. Conversely, the trading bourgeoisies and their class allies generally pursued the same policy inherited from the colonial period, a liberal system which made no attempt to deck itself in an appearance of socialism.

The petty bourgeoisie, especially when it succeeded in mobilizing alongside it the broad masses on a radical platform, generally tried to secure a base for its regime, by, among other things, taking external trade away from the traders, nationalizing the economic interests of big capital, taking a whole series of measures which enabled it to claim to be socialist.

But historical experience has proved that, after a few years of management which amounted virtually to a private management of public property, it was transformed into a true bureaucratic bourgeoisie whose economic base was built up from control of the state and public property.

Like every bourgeoisie in an underdeveloped country faced with the omnipotence of big capital, it is condemned, in order to consolidate itself, to strengthen these links with the world market. Thus, the bureaucratic bourgeoisie always ends up entering the order of the world capitalist system just like any other bourgeoisie. Big capital is all the more ready to make a few concessions to it on statizations because it knows that it can always extract enormous profits from the countries controlled by the bureaucracies, and acquire economic positions out of all proportion to the slight concessions made.

Thus, in the long run, the bureaucratic bourgeoisie and the other commercial or agrarian bourgeoisies always end up looking for a deal, their interests coinciding in the alliance with big capital.

We feel that what has just been described can be said of all African bourgeoisies, and of their composition and sometimes of the way in which they came to be formed. Today they control most African states and manage them in accordance with their common interests.

The increasingly common interventions by the military on the political stage can only be interpreted as arbitrating within the bourgeois classes, or between the bourgeois classes and the radical intellectual petty bourgeoisie. These armies can neither have nor offer new solutions. Their political orientations cannot but be inspired by the lines followed either by the classic African bourgeoisies or by the radical bourgeoisies in transition.

In fact, the most important thing about African armies that seize power is that, in general, they are the product of the colonial system. They are linked either to the African bourgeoisies, or to the radical petty bourgeoisies. In any event, the army as a social group linked to a given class can neither imagine nor have a political line of its own that can be any different.

If we turn now to the broad masses, mainly the peasantry and the working class but also the middle strata that might join it, it cannot be said that since independence consciousness of common interests has strengthened as it has among the bourgeois strata and classes.

They are usually not organized on the basis of their own well-perceived interests. The result is that they act as a force to be manoeuvred and manipulated by ruling political parties that are generally bourgeois in nature. The political organizations that exist in this or that African country (usually, they are only single parties with every possible political language and contradictory political practices) and that claim to be Marxist and try to mobilize the masses so that they can liberate themselves and assume the historic tasks that fall on them are only tiny organizations with a very weak capacity for mobilization because of the petty bourgeois intellectual origins of the militants in these organizations.

The classes in power really have no interest in seeing the masses organize themselves freely and develop their political consciousness clearly. It follows that the organizations that represent them officially are, in fact, simply links in the state structures.

And yet, peasantries, working classes, middle classes and petty bourgeois strata represent more than 97% of the population in most African countries. In many countries, the peasantry alone represents 80% of the adult working population. The characteristics of this key stratum naturally vary in detail from country to country, but it has many common traits.

If we leave aside the rural bourgeoisie that is very small in most countries, the differentiations among the peasants themselves are generally not very significant. The most typical feature of the African peasantry seems to be its excessive dispersal in often tiny villages. The consequence of this is that disseminating ideas among the peasants is not easy, which makes it difficult to bring them together in an organization independent of the political authorities and fighting for their own cause.

From that flows the lack of concern on the part of the authorities for their cause. This shows itself in illiteracy and in the inadequacy, even the total absence, of infrastructure in the rural areas. The peasants, the main producers of agricultural products, are the first ones to be affected by famine, malnutrition and poverty. It follows that a radical transformation in an African state ought to take as its starting point a radical change in the conditions of existence in the countryside. And that would require that the peasants be able to come together freely in a solid organization which would be a powerful force of dialogue and decision-making on all the great problems of the nation.

As for the working class, it is only really of any size in a tiny number of African countries. Where it is somewhat organized, their organizations do not escape the rule that in most countries all organizations are subordinate to political parties, to government: the trade unions and particularly their leadership, made up of aristocrats whose job is to muzzle the workers by stifling any incipient sign of combativeness, do not play their role.

As for the middle classes—the clerks, artisans, etc.—sometimes they are organized in trade unions that are independent of the workers' unions, sometimes they are simply integrated into these unions. Their possibilities of struggling are thus no greater.

Only the small intellectual unions of teachers or white-collar workers are frequently active and embark on large-scale actions against governments for their clearly understood interests. But without the active support of the rest of the mass of workers, their capacity for struggle has difficulty pushing the dominant structures in the directions of real and profound change.

Such, schematically, has been the evolution of class structures, types of organizations and forms of struggle in most African states since independence. It gives little ground for hope that, in the near future, there will be any marked improvement. But nor should it lead to total pessimism. The potential forces for change exist. They need to be able to act so as to take in hand their own fates and benefit from the fruit of their creative labour.

What popular alliance for the alternative?

In order to reply to this question, we need to start from a simple observation: the incapacity of most ruling political classes to undertake a policy which would overcome the crisis and guarantee decent living conditions for all strata of the population. If such is the situation, it is not without reason. We think that it is essentially due to the level of consumption of these ruling classes and their vast expenditure which lead to a waste of national resources and the massive impoverishment of the vast majority of the population.

The policy that they have hitherto followed has not made possible national accumulation for investment to put all the nation's manpower to work. What is produced by those who work is diverted to minority interests whose national character is questionable. Only the popular strata can implement a national popular programme by putting an end to useless expenditure of dubious national interest.

In our opinion, these popular strata in most countries in Africa would include the peasantry (farmers, herders, fishermen), the working class, the intermediate classes, that is, employees, artisans, small traders, the petty bourgeoisie of officials in the administration, teaching, etc. The interests of these classes and strata are not always identical. But they can come together in the phase of constructing a national popular programme.

We believe that every possible pressure from all national and international agencies must be brought to bear for these strata to secure freedom of organization. This would be for them the first step to having real access to the structures of power. They will necessarily transform these structures since these transformations—transformations in depth—will be the condition for true and visible improvement in their own conditions of existence and, by the same token, the improvement of the conditions of existence of all strata of the population.

On the basis of their common interests, these popular strata could constitute a popular alliance bringing together their completely free autonomous organizations. They are, we feel, the only ones likely to implement a national popular programme which could overcome the crisis. That naturally presupposes that they control every structure of power.

What national popular programme might the popular alliance propose?

Although our work is mainly focused on agriculture, it is clear that an alternative to the crisis cannot rest solely on an agricultural programme. It must of necessity take into account all aspects of economic and social life. Thus, while stressing agricultural problems, we shall relate them to other economic sectors without whose transformation agriculture has no hope of truly developing to ensure food self-sufficiency for African countries. We have stressed sufficiently the failure of the development strategies hitherto pursued in the framework of the existing socio-political systems. Proposing a national popular programme that could constitute an alternative to these strategies can only emerge from a sustained work of reflection which must be the constant preoccupation of researchers, especially African researchers.

It would seem that among the political powers that be some are beginning increasingly to see the necessity of an alternative. The question is: will they be in a position to promote such a programme, given the powerful dominant interests that will oppose it? Whatever the case may be, in the Lagos Plan of Action adopted at the OAU's first economic summit in Lagos in April 1980, the OAU included some phrases that give the impression that it is increasingly realized that only solutions aimed at the real satisfaction of the basic needs of the broad masses can solve the serious crises affecting Africa, which are only intensifying.

Thus, in chapter I of this programme, devoted to Food and Agriculture, it is stated: 'At the root of the food problem in Africa is the fact that Governments have not usually accorded the necessary priority to agriculture both in the allocation of resources, and in giving sufficient attention to policies for the promotion of productivity and improvement of rural life.' Such an admission is an advance over the usual blame attributed to natural disasters and the deterioration of the terms of trade, although of course these do play a part in the crisis as aggravating factors.

Again, in the introduction to the document, stress is repeatedly laid on the notion that Africa must cultivate the virtue of self-sufficiency. But there is a big difference between statements of principles and all-embracing policies designed to realize them, and one may doubt whether the move from the one to the other is possible in the socio-political conditions currently existing in most African states.

In the Lagos document, technical measures are proposed in all areas of economic life to achieve this self-sufficiency. But since the political conditions that would make the application of these measures possible are not mentioned, there is an obvious risk of remaining once again at the level of statements of good intentions. That being the case, when it comes to the content to be given to a national popular programme in the event of the broad masses holding the reins of political power, one can do no more than offer reflections since here too there is the risk of lapsing into utopianism by wanting to go into detail about the reorganization of a concrete society, starting from abstract *a priori* assumptions. But these reflections on a popular programme should make it possible to put forward a few general ideas which would be worth analysing further and eventually challenging through research, failing any possibility of concrete experimentation.

Thus, without falling into a pragmatism that totally ignores theoretical achievements capable of helping in the elaboration of a comprehensive policy directed towards the interests of the broad masses, we feel that experimentation and trial and error will play a not insignificant role in building up such a programme to bring closer socio-economic measures which quite often seem to be ruled out because of the whole heritage of the past. We shall thus be obliged to spend more time on the scale of the problems that will have to be faced than on the formulation of ready-made solutions.

Starting from this careful approach, we think that the first problem that ought to be examined—and which is both economic and political—is that of borders. We shall formulate it in this way: given that the political prerequisites, that is, the alteration of the internal relations of domination that exist in the various countries, were achieved, is it possible in the tiny confines of most African states to implement a national popular programme that calls for delinking from the world capitalist system? Explicitly, would it be possible for a small African state that fundamentally altered its socio-political structures to implement a comprehensive economic policy of independence breaking with the system and sustain this policy over time? What is quite clear is that the pressures from imperialism and other African states, especially neighbouring ones linked to imperialism, will be very strong. It will need a powerful mobilization of the entire working people of this country to be able to stand up to it.

The disappearance of borders, which are usually artificial creations corresponding to the interests of the colonial economy, is both an economic and a political necessity. The intra-African trade patterns that existed before the imposition of colonialism were disrupted and replaced by trade oriented towards the metropolis. Railways, ports and even the development of cities, were conceived in this perspective, leading to an unequal development between seriously underdeveloped countries.

In order to restructure the economies of most countries, borders that correspond to no socio-historical, or even ethnic and geographical, reality would have to disappear, even if gradually, and give way to viable economic units capable of resisting the assaults of imperialism. Of course, it is true that imperialism too sometimes makes similar suggestions, but it does so for its own aims designed to strengthen the dependency of several states brought together under its sway.

The creation of large economic zones transcending borders that should be challenged can therefore only be a popular achievement if it is carried out by alliances among popular strata which truly and effectively suppress artificial and anachronistic barriers that do not reflect the history of their people, and which would be brakes on the implementation of popular programmes in their respective countries. This would make possible the creation, between states embarked on a strategy of delinking from the world capitalist system, of the conditions for the implementation of a unified and coherent development plan, taking account of the economic potential of the various countries and their necessary complementarity. Thus, the way would be open to solid intra-African economic links and a collective autocentred autonomy.

The disappearance of the borders inherited from the colonial system can, in the framework of an alternative to the crisis, open up far-ranging perspectives for the economic development of Africa. But the OAU, as it exists, does not exhibit the least sign of embarking on this path because of the major contradictions running through it and the dominant class interests of the states belonging to it. Thus, the future of Africa to attain an autocentred collective autonomy of development can only be the work of popular alliances effectively controlling all powers, and on the lines that we have sketched out above.

This real unification, as opposed to unification that is simply a matter of diplomatic proclamations, would make possible the unification of African producers to face the demand of the market for the products that Africa is obliged to sell on that market. Indeed, delinking with the world capitalist system can in no way mean a total break with this system. It means that the socio-economic structures of Africa would no longer be organized and oriented in terms of the interests of this system as is currently the case, but in terms of the interests of the toiling African masses. It is through this true unification of producers that Africa would be able to have a considerable influence on the world prices of the raw materials that it supplies until it is able to realize the goal sought, which is the on-the-spot processing of these raw materials to meet the needs of African economies.

It is because they can provoke isolation and rivalry among countries producing cocoa, coffee, iron ore, uranium etc., that the developed countries control price-fixing on the world market. The existing governments, because of these rivalries, their attachment to personal power or even the vulnerability of many of them to corruption, simply play into the hands of the developed capitalist countries and the imperialist monopolies.

By adopting a monopoly situation for the various African commodities popular governments could, if not reverse the direction of dependence in Africa's favour, at least create a degree of balance between the various parties in the world market. African products could then be sold at prices close to their value, that is, taking account of the quantities of labour incorporated into them, which return to the labour provided in Africa its value, and taking account of all the factors that come into the fixing of prices as if these goods were produced in the developed capitalist countries.

The union of producers faced with the demands of the world market will only be fully effective if external trade—commodity imports and exports—as well as the import of capital are conducted exclusively by the popular government. That is essential in order to avoid big capital becoming involved in the economies by controlling important sectors.

With these basic conditions set down, there is still a need to reflect on the manner of reorganizing production and exchanges among the various sectors of the economy to achieve self-reliance, the immediate aim of which is food self-sufficiency.

In approaching this question, it cannot be forgotten that there are priorities to be realized and which are realizable once the political prerequisites have been realized. There is a need to:

— Raise agricultural production and productivity substantially and rapidly, with priority for food production.

— Ensure that the increase in agricultural production is accompanied by a substantial raising of the standard of living of the peasants in the countryside.

— Promote a selective industrialization coordinated with and sustaining the development of agriculture.

— Give the manpower in these industries wage levels that guarantee the full reproduction of their labour power and the upkeep of their family.

— Produce mainly for the domestic market in both agriculture and industry and be oriented wholly towards mass production.

It goes without saying that these, the most pressing, measures must be accompanied by other complementary measures guaranteeing their success. They include:

— Putting an end to the rural exodus and reducing urbanization to a level low enough not to hinder economic and social policy overall. This can be achieved by creating the conditions for a massive voluntary return to the countryside where development policy will be concentrated far more than in the towns.

— Completely transforming teaching and education to adapt them to new forms of development and to make them respond to these needs.

It would then be necessary to reflect on the sectoral and overall policies that would be needed to carry out these first urgent measures rapidly.

The rural areas

The new economic and social policy must be centred mainly on the development of the rural areas which have hitherto been bled white to be able to carry out policies linked to the interests of a few possessing classes and hence unpopular policies.

The starting point for a policy to develop the rural areas seems to lie in the peasants and the rural areas as a whole having the possibility of organizing themselves freely and independently. These organizations which would be created all over the rural areas must be able to enjoy full decision-making powers. The basic cell of these organisations in the rural areas might be the village, as Guy Belloncle suggests in his book *La Question Paysanne en Afrique Noire* (published by Editions Karthala).

However, it is important not to over-romanticize the traditional village structures. They sometimes involve ideologically-backed social relations that might constitute brakes on development. In fact, in most countries at the village level there exist lineage, tribal or caste relations solidly sustained by fossilized ideologies hostile to progress.

These outmoded relations must be undermined by organization and work; they often involve disguised or visible links of dependency that obscure and restrict 'democratic power structures', about which Guy Belloncle writes in the book mentioned above. It is then that true village communities could come into being shorn of age-old defects where, with the equality of all in decision-making eliminating the hierarchy of established authorities, they would reinforce the solidarity of all in the community as a whole. An organization thus built from the village upwards and spread to the whole countryside with strong structures could constitute the backbone of the new governments in countries where the rural areas are so dominant.

Being thus organized at the national level the peasantry, or more precisely the rural inhabitants, would actively participate in the preparation of the economic and social policy of the new governments and in supervising its implementation. The rural areas would be present at the highest level of the new political system through their leading bodies and their delegates, and at the base through the whole village community.

In many countries this reorganization of the rural areas would necessitate the amalgamation of numerous villages that are too small and too scattered. Through this amalgamation of villages to achieve a relatively large size, it is possible to realize a number of goals, the most important of which is certainly to shift peasant life and work from the narrow family framework to the village communal framework. At one stroke this would facilitate the circulation of information, be it political or technical. More egalitarian access to inputs and the means of production would be promoted. Access to the infrastructure of schools, health facilities and other basic services would be achieved much more cheaply.

The countryside must offer every facility to guarantee the total security of the rural worker and his family throughout their lives in the framework of the village community.

How can this be achieved on the economic level? We do not think there is any need to have recourse to vast sums of capital coming from abroad, which are swallowed up in endless studies for which the peasant has to pay but whose results he never sees.

Use must first be made of the immense resources of the rural areas, hitherto wasted, pillaged, misappropriated, taxed and transferred to the benefit of other social strata.

Access to land: First, access to land must be secured to all and be equitable for all. In countries where big landowners exist, all land will purely and simply be transferred to the village communities which will be responsible for their fair allocation while moving towards their collective development.

For land that requires irrigation, the labour could be performed by human investment by the village community, aided technically, where necessary, by the governing organs of popular power.

What must be avoided at all costs is big impressive dams built at the cost of millions of dollars borrowed from abroad, repayment of which mortgages national economic policy and is beyond peasant resources.

Production: On the level of production, with nothing ruled out, every product that can be a source of food must be sought out and developed: crop products and products of gathering. Research will be necessary into the conservation or processing of some of them. The fact is that it is not uncommon in the African bush and forest to see fruit trees bear their whole crop at certain seasons. These perfectly edible fruits finish up going rotten for lack of means of picking and preserving them.

Added to this is the tendency in some fairly well-off strata—except in periods of famine, of course—to consider some products as non-'noble', products of gathering in particular, whereas they may be very rich in calories and vitamins. There is work to be done to sensitize and to inform people so as to turn these into products products of mass consumption, which would make it possible to enrich and vary diets.

As for cash crops not intended for consumption, like cotton, or which are not for everyday consumption, like coffee and cocoa, there can be no question of halting their production since they constitute important resources for countries and peoples. There will need to be studies of how to regulate production of them in terms first of the food needs of the population, and then of the demands of the international market.

Inputs and improving techniques: This is one of the most difficult problems to solve since the inputs are not, generally, produced on the spot and their cost on the world market is high and rising sharply from year to year. Imports

system. Thus, it will facilitate and ensure balance in the exchanges between producers at the local level. In the same way, it will be the sole intermediary in exchanges between the village and the world outside the village, especially exchanges with the Co-operative Union. This will purchase its products and sell it the inputs and other products that it needs. A profit margin will be allowed the co-operative. Its profits will go into a loan account and be used for public works in the village community.

It must be laid down that the main aim of the loan account is to strengthen solidarity within the community. Its prime purpose will thus be to help its members in difficulties: peasants who lack machinery, herdsmen needing watering troughs, etc.

The co-operative, with a mainly economic function, will be under the control of the village assembly which will have a much more political function. The village assembly will be able at any time to check the management of the co-operative and it will transmit directives to it to meet the needs of the community. In addition, the assembly will undertake and sustain the political, economic and social mobilization of the community. Gradually, it will seek the means to move from family to communal labour and farming in more and more areas.

The urban areas

The strategy for the urban areas will be closely related to the rural development strategy, so that the two strategies are mutually reinforcing for a harmonious and balanced development.

To date, industrialization has affected only the urban centres in most countries. That encourages the rural exodus, even though the supply of industrial jobs is low and in no way corresponds to the numbers involved in the exodus. That is due to the policies which are currently being pursued and which manifestly lack imagination.

What is needed in most cases is to devise the conditions for a movement in the opposite direction. What is needed is to initiate and encourage an urban exodus so as to reduce considerably the population of the towns. A well thought out policy with a variety of axes might make it possible to attain that objective. A complete decentralization of the industries currently concentrated in the towns, sometimes even in the capital alone, could be a major axis.

The goal of industrial production must be twofold. First, the production of agricultural inputs through a host of small local industries scattered all over the country. Second, the processing of the raw materials which an expanding agriculture supported in that way could produce to ensure that these industries run at full capacity.

Without going into too much detail, it is reasonable to think that the small production industries could, for technical matters and management, come under the basic structures while the relatively heavier processing industries could mainly come under the central government. Thus, processing industries, like mining industries, would be nationally owned, whereas small industries which will initially include a not insignificant amount of crafts, would be owned by the basic structures.

The creation in the rural communities, of vital infrastructure (in education, health, sport, etc.), linked to the development of small rural industries might halt and even reverse the rural exodus. The great mass of the urban unemployed of rural origin, who suffer from malnutrition and poor health, would surely return to the countryside where they would henceforth be guaranteed a job and all the material and psychological conditions that provide a family with security. In addition, the costs of building this rural infrastructure would be very low, far less than the costs of the excessive luxury that marks some sectors in most of our cities.

The marginalized urban craft sectors could be grouped together and properly structured in order to meet urban demand efficiently, and so greatly reduce imports. We are thinking, among others, of the construction sector, furniture manufacture, etc. They could thus constitute a real industry and guarantee regular work to those working there.

The central authorities have an important role to play in the reorganization of these sectors and in facilitating the acquisition of some machinery.

The activities of countless petty traders which often conceal hidden unemployment must be rethought. Increasing the number of people's shops in the towns catering for all everyday consumer products might make it possible to recycle a good proportion of these semi-employed people.

The administration should be as decentralized as possible. In addition, to avoid the plethora of officials barricaded in their offices in towns, the teaching profession must be entirely reformed, eliminating all the sequels of the colonial period so as to adapt it to the new economic and social realities.

The training of cadres at all levels cannot remain unorganized. It must be strictly planned so as to respond to the needs of the rural and urban economies and the services essential to them. Thus, teaching and training will be mainly directed towards the productive system.

Organizing the suggestions

This problem is the key one and one that is difficult to resolve because of the numerous dangers to be avoided; risk of bureaucratization, lack of congruence between decision-making centres, decisions emanating from the base structures, and provisions adopted by the central authorities of the popular alliances. We shall simply put forward a few thoughts. First, the question may be asked whether the plan will be simply indicative (in which case the state intervenes in the economic arena in only a limited way), or binding (the state then has a very extensive field of action). We think that the strategies sketched out can only be developed in the framework of a binding plan. This binding plan must, however, be sufficiently flexible for the base structures to retain considerable initiative. But it is absolutely necessary that the state established by the popular alliances intervene and coordinate minutely all economic activities: budget, investment, income distribution, wages, prices, etc. This is necessary so that all social strata may receive fair rewards for their labour and so that all can enjoy their full rights. This plan should aim to reduce considerably the capitalist sectors that would remain in this transitional economy, and subject them to strict control.

The problem of employment and the fixing of prices and wages must be settled by the central bodies in terms of the necessary transformation of the structures of the economy and improving the living conditions of the whole population. At the relatively primary stage reached by most African economies, there does not seem to be any need for complex models based on very detailed economic calculations in order to prepare a completely coherent plan taking account of the various imperatives. Trial and error and intuition will still have an important role to play in the success of a plan which, to repeat, cannot but be holistic and centralized, contrary to the catalogues of projects that are currently presented in most countries as being plans.

For its financing, the plan must rely predominantly on domestic resources. This is necessary to ensure economic independence and will be made possible through the systematic elimination of waste of all sorts, and the end of prestige expenditure that is as useless as it is costly. Given the high level of corruption among some highly paid strata in Africa, there will certainly be reason to investigate fortunes, especially large fortunes, to retrieve illicitly acquired gains for the benefit of the nation and the people. That said, the question of financing the plan remains to be solved. Given the nationalization of external trade and the steep reduction of imports, customs receipts will be very severely limited.

We believe that the investment funds should come mainly from taxes paid by state and private enterprises and from a wealth tax. A subsidiary source will also be taxes paid by workers based on their income and the standard of living to be guaranteed them. In addition, funds could come from income from exports of resources that would have to be exported, either because they could not yet be processed on the spot or because their supply exceeds domestic needs.

Finally, to encourage and sustain the sectors vital for the transformation of structures and begin true development, the fixing of prices and wages will play an important role both in allocating manpower suitably and in controlling consumption.

To summarize, we have here simply put forward a few ideas that are the conditions of, and may make possible, a true alternative to the present crisis situation. It all calls for a vast effort of organization and thought. It goes without saying that success can only be assured by a strong mobilization of the mass of workers freed from exploitation and oppression, and giving free rein to their creative initiative.

Conclusion

It is not only the African peoples who perceive the seriousness of the crisis because they are experiencing it. There are also the promoters of the system who see in it the threat to their survival.

If, for the African peoples, possible solutions emerge from class struggles internal to their societies, for the defenders of the system the first thing is to safeguard it by patching it up and adjusting it a bit here and there to make the consequences of the crisis more bearable. But this adjusting and patching run up against the constraints inherent in the system.

That is why the new and fashionable analyses, the new formulations in vogue (new international economic order, North-South dialogue), developed by experts used to this type of exercise, always end up getting nowhere because the sacrosanct law of profit is there. And it is this law that ultimately determines what is acceptable to the system.

Adjustments within the system, however they are described, cannot provide any solution to the poverty of the dominated peoples.

The fact is that the capitalist system rests on class relations and hence on relations of force. So long as the dominated peoples do not constitute a true force by their unity, they will only be listened to out of simple politeness.

As for the African peoples, they will only find salvation the day they build the Africa of the peoples. Unfortunately, we do not seem to be getting much nearer to that day.

Appendix: Complaints of the Rice-growers of San (Mali)

Report on the alarming situation in field B of San-West

We, rice-growers of San, have the honour to report as follows:

Since the 1971-72 agricultural season, we have been suffering from drought, in the framework of rice-growing, a phenomenon which is by no means peculiar to our *cercle*. Happily our cries of alarm were heard by the authorities at all levels. Thus, the government laid out the San-West plain in two fields, A and B, and established a pumping system there, a sensational occurrence which was joyfully welcomed at the time by all the rice-growers. This privileged situation gave us the occasion to hold several general meetings during which numerous expressions of gratitude and thanks were addressed to the CMLN and the government. And we were already thinking that in no time at all most of our hopes would be realized. But alas! Our enthusiasm soon gave way to discouragement, a situation brought about by the procedure for working the irrigated perimeter. In fact, the conditions demanded by Operation Rice did not allow us, despite three years of strenuous efforts, to obtain the grain necessary for the annual subsistence of our families. Thus, in field A, the quota due is 900 kg. of paddy rice per plot of 1.5 ha., or 600 kg. per ha. In B, it is 400 kg./ha. In addition, we commercialize: field A, 800 kg./ha. This means that virtually the entire harvest of the field goes for the quota due or commercialization. The little that is left over, two, three, four, five or six sacks of paddy, goes to the rice-grower. Of course, he receives a small sum (after the value of all the credits that he has enjoyed from Operation Rice—fertilizer, loan of the seeder, etc.—have been deducted), and collects it sulking impotently. But while this peasant possesses at least 3 hectares in field B, it is the whole harvest of field A that is commercialized.

As for field B, where production is lower because of the ageing of the soil and the presence of *diga* [a weed], the peasant often pays the whole or the balance of his quota in kind, often borrowed, a phenomenon which in our opinion seems to contradict the objectives of the CMLN and the government, whose aim is to help the peasantry. However, during the general meetings we have more than once denounced this practice which starves and impoverishes us more each year, but the officials responsible for Operation Rice have always taken refuge behind the assertion that 'the demands imposed on rice-growers are the instructions we receive from our leaders.' And the peasants' response has always been to point out to the competent authorities the evil character of such a practice. This misunderstanding provoked such discontent that the rice-growers suggested to the officials of Operation Rice: 'If our complaints are not examined, it would be better to stop convening general meetings, and simply publish decisions through public notices.'

To calm people down, we were promised that caterpillar tractors would be brought in, five (5) in number, to fight against the *diga*, and that this would take place by the end of the 1977–78 season. Not being sure that the said machines would come, we suggested doing the low water ploughing with a plough as we usually did, but we constantly met with the refusal of the Field Leader who even threatened peasants who might persist in doing so. And all folded their arms and waited for the end of the ploughing that was supposed to be done by the caterpillar tractors. Unfortunately, we received only two machines which, as at 5 April 1978, had not ploughed more than 100 out of 700 ha. Faced with their inability to keep their promise, the Operation Rice officials ordered what they had always rejected, that is, using the plough. Now, at this time, the state of the soil was such that even the tractor had difficulty in doing effective ploughing, and how much more so the plough. The general meeting stressed that responsibility for this unhappy situation fell wholly on the cadres of Operation Rice.

The tornado of 28 April 1978, which saw 132 mm. of rainfall, made the *diga* grow but our oxen were so weakened by the meagre pasturage that we could not continue with the ploughing.

At the general meeting on 8 June 1978, we were asked to plough and sow fields A and B where the caterpillar tractor had carried out ploughing between 15 and 30 June. All the farmers in field A carried out their work with care since low water ploughing had been done there. But in B, despite all our goodwill and all the efforts made, the state of the soil did not allow us to record the same results. Nevertheless, the few peasants who had been able to plough some land did not unfortunately obtain the seeds they had asked for. They had to face new demands from the Field Leader which were: to complete entirely the ploughing and weeding in order to be able to claim the seeds that we were to buy; to sow all the ploughed areas in drills, contrary to the practice of broadcast sowing that we had previously applied.

Despite the insistence of the rice-growers, the Field Leader remained deaf to every explanation trying to make him see reason. Thus, the long period between 28 April and 25 June should have enabled a large number of peasants to replough and sow if not all, at least 75% of field B if low-water ploughing with the plough had been authorized from the outset.

To explain his demands, the Field Leader dared to claim that he was worried that the rice farmers would consume the seed if it was distributed to them before the end of the weeding. This concern on the part of the Field Leader, expressed out loud, provoked indignation among the peasants at the general meeting on 21 July 1978.

In fact, the said general meeting was held when Operation Rice, completely thrown out of gear by the unfortunate consequences of his obstinacy, made the administrative and municipal authorities aware of the scale of the disaster due to the rains which had wholly submerged field B, making it thus uncultivable for the current season. Here is a breakdown of the rainfall recorded during the period concerned:

3 July	35.2 mm.	28 July	33.8 mm.
8 July	28.4 mm.	29 July	9.5 mm.
11 July	60.3 mm.	30 July	0.9 mm.
14 July	53.8 mm.	31 July	15.9 mm.
15 July	6.2 mm.	1 Aug	2.9 mm.
18 July	23.8 mm.	2 Aug	23.7 mm.
19 July	0.3 mm.	3 Aug	3.8 mm.
23 July	21.6 mm.	8 Aug	3.3 mm.
26 July	trace	11 Aug	27.2 mm.
		Total	350.0 mm.

To try and make up for their mistakes, we were advised during the general meeting of 21 July 1978 to employ every possible means to do something. And to save face recourse was had to the following procedure: hurried distribution of seeds, a system Operation Rice had previously condemned; permission to use broadcast sowing.

So everybody went to work with more ardour. But no one can perform the impossible. The height of the grass and the amount of water and mud in which we were slopping around discouraged even the most determined. That is why we are now launching an urgent appeal to all the local, regional and national authorities to examine with special attention the situation of field B of San-West, which we consider a disaster. We have no food left from the 1977–78 harvest and no longer expect anything from the current season. We are therefore putting all our hope in the CMLN and the Government to come and help us.

Given all the above, we find ourselves obliged to submit for examination by the authorities some of the complaints that we deem urgent:

(1) We urgently request that the population be supplied with cereals and that seeds be set aside for the next season.

(2) We ask for the laying out in the near future of lower field A which, in our view, could be as profitable as the previous one.

(3) We recommend that an effective struggle be waged against *diga* in field B with a view to making it worth sowing, and that this should be done immediately the waters recede.

(4) We draw the attention of the competent authorities to the fact that virtually all our harvest is taken up with the quota, threshing costs and commercialization, and we ask that these burdens be lightened.

Our aim is to alert all the authorities so that they help us to save our sinking boat.

Copies to:

CMLN	(5)
Prime Minister	(2)
Min. Rural Dev.	(1)
Sec. of State for Industry	(1)
Regional Governor	(1)
Commandant de cercle, San	(2)
San Branch UDPM Bureau	(2)
Mayor Commune San	(2)
Operation Rice San	(1)
Archives of Board of Management	(1)

Chairman of the Session
Allay Daou

Secretary of the Session
Karamoko Traore

San, 15 August 1978

Select Bibliography

Africa Development, Various issues (Codesria, Dakar).

Amidu, Magasa. *Papa Commandant a Jeté un Grand Filet devant Nous* (Editions Maspero, Paris, 1978).

Amin, Samir. *L'Accumulation à l'Echelle Mondiale* (Editions Anthropos, 1971); Eng. translation, *Accumulation on a World Scale* (Monthly Review Press, New York, 1975).

——— *Le Développement Inégal* (Editions de Minuit, Paris, 1975); Eng. translation, *Unequal Development* (Harvester Press, Hassocks, 1976).

Amin, Samir, and others. *L'Agriculture et le Capitalisme en Afrique* (Editions Anthropos—IDEP, 1975).

Amin, Samir, Alexandre Faire and Daniel Malkin. *L'Avenir Industriel de l'Afrique*. (Editions Harmattan—ACCT, 1980).

Amin, S. *Neo-colonialism in West Africa* (Penguin, Harmondsworth, 1973)

Balandier, G. *Sociologie Actuelle de l'Afrique Noire* (Editions PUF, 1955).

Belloncle, Guy. *Coopérative et Développement en Afrique Noire* (CEDEC, 1979).

——— *Quel Développement Rural pour l'Afrique Noire?* (Nouvelles Editions Africaines, 1979).

——— *La Question Paysanne en Afrique Noire* (Editions Karthala, 1982).

Benot, Y. *Idéologies des Indépendances Africaines* (Editions Maspero, Paris, 1969).

Brugnicourt, J. *Les Marginaux de l'Economie Moderne dans le Sahel* (Sahel Hebdo, 1975).

Cabral, A. *Unité et Lutte* (Maspero, Paris, 1975); Eng. translation, *Unity and Struggle* (London, 1980).

Cahiers du C.E.S. *L'Afrique Recolonisée* (C.C.E.S., 1966).

C.E.R.M. *Sur le Mode de Production Asiatique* (Editions Sociales, 1970).

C.E.R.M. *Sur les Sociétés Précapitalistes* (Editions Sociales, 1970).

Chenery, H., and others. *Redistribution with Growth* (Oxford University Press, London, 1977).

Cissé, Moussa, and others. *Mali: le Paysan et l'Etat*. (Editions Harmattan, 1981).

Cissé, Y., and Emile Leynaud. *Paysannerie Malinké du Haut Niger* (Editions Imprimeries Populaires du Mali, 1978).

Collins, Joseph and Frances Moore Lappe. *L'Industrie de la Faim* (Editions de l'Etincelle, 1977).

Comité Catholique. *Contre la Faim et pour le Développement* (1979).

Comité Information, Sahel. *Qui se Nourrit de la Famine en Afrique?* (Editions Maspero, 1974).

Coquery-Vidrovitch, C. 'L'Afrique et la Crise de 1930', Colloquium at the University of Paris VII, 9-10 April 1976.

——— *De l'Impérialisme Ancien à l'Impérialisme Moderne: l'Avatar Colonial*, in IDEP Reproduction/243.

Critique de l'Economie Politique, no.15, *Paysannerie et Réformes Agraires* (Editions Maspero).

Dembélé, Kari. 'Sociologie des Paysans Maliens,' unpublished 3rd cycle thesis, University of Paris-Nanterre, 1970.

Derriennic, Hervé. *Famines et Domination en Afrique Noire.* (Editions Harmattan, 1977).

Diagne, P. *Pour l'Unité Ouest-Africaine: Micro-Etats et Intégration Economique* (Editions Anthropos, 1972).

Dieng, A. 'Formations Sociologiques et Commerce à Longue Distance en Afrique de l'Ouest,' IDEP Reproduction/385.

Diop, Majhmout. *Histoire et Classes Sociales dans l'Afrique de l'Ouest:* vol. 1, *Le Mali* (Editions Maspero, 1971); vol. 2, *Le Sénégal* (Editions Maspero, 1972).

Dumont, René. *L'Afrique Noire est Mal Partie* (Editions du Seuil, Paris, 1969); Eng. translation, *False Start in Africa*

——— *Paysanneries aux Abois* (Editions du Seuil, Paris, 1972).

——— *La Croissance de la Famine* (Editions du Seuil, Paris, 1981).

Economie Rurale en Chine. Peking Publishing House, January, 1976.

Engels, F. *L'Origine de la Famille, de la Propriété Privée et de l'Etat* (Editions Sociales, 1966); Eng. edition, *The Origin of the Family, Private Property and the State* (Progress Publishers, Moscow, 1969).

Fabre, R. *Economie du Développement* (C.E.S.D., Paris, 1967).

Faure, Claude. 'Sur les Formes de Domination du Capital,' Department of Political Economy, University of Paris VIII, Vincennes.

Feder, Ernest. 'Macnamara's Little Green Revolution,' *Economic and Political Weekly*, April 1976.

Food and Agriculture Organization of the UN. Series of documents.

Founou, Bernard T. 'Marché réel et marché formel de la force de travail', *La Pensée*, August-September 1974.

——— 'L'approche par les besoins essentiels, une théorie nouvelle de développement pour la décennie 80?', *Africa Development*, 1980.

——— *Fondements de l'Economie de Traite au Sénégal* (Editions Silex, 1981).

Frank, André Gunder. 'The Development of Underdevelopment,' in *Latin, America: Underdevelopment or Revolution* (Monthly Review Press, New York And London, 1969).

George, Susan. *How the Other Half Dies* (Penguin Books, Harmondsworth, 1976).

Gutelman, Michel. *L'Agriculture Socialisée à Cuba* (Editions Maspero, 1967).

International Labour Organisation (ILO). *L'Animation Rurale dans les Pays en Afrique Francophone* (ILO, Geneva, 1970).

——— *Employment, Growth and Basic Needs* (ILO, Geneva, 1976).

Lele, Uma. *The Design for Rural Development* (Johns Hopkins University Press for World Bank, Washington D.C., 1975).

Lenin, V.I. *Imperialism: the Highest Stage of Capitalism* (1917).

——— *The Development of Capitalism in Russia* (Progress Publishers, Moscow, 1977).

Marx, K. *A Contribution to the Critique of Political Economy* (Lawrence and Wishart, London, 1971).

Meillassoux, Claude, *Femmes, Greniers et Capitaux* (Editions Maspero, Paris, 1977).
Mendras, H., and Y. Tavernier. *Terre, Paysans et Politique*, 3 vols. (SEDEIS, Paris, 1969).
Nkrumah, K. *Africa Must Unite* (Heinemann, London, 1963).
——— *Neocolonialism, the Highest Stage of Imperialism* (Heinemann, London, 1965).
——— *The Class Struggle in Africa* (1970).
Olivier, J.P. *Afrique, qui exploite qui?, Temps Modernes*, May-June 1975.
Opération Arachide et Culture Vivrière (Groundnut and Foodcrop Programme) (Mali, 1978).
Operation Riz-Ségou (Segou Rice Project) (Mali, June 1977).
Organisation for Economic Cooperation and Development (OECD). *Studies of Trends in World Supply and Demand of Major Agricultural Commodities* (OECD, Paris, 1976).
Pélissier, P. *Les Paysans du Sénégal* (Imprimerie Fabriques St Yriex, 1966).
Projet de développement rural de Sedhiou, Sénégal (Sedhiou Rural Development Project) (Senegal, 1976-77).
Provent, A., and F. de Ravignan. *Le Nouvel Ordre de la Faim* (Editions du Seuil, Paris, 1977).
Rey, P.-P. 'Le Transfert de Surtravail de la Paysannerie vers le Capitalisme: réponse à l'article de James Becker, "On the Monopoly Theory of Monopoly Capitalism,"' in *Science and Society*, 1970.
Suret-Canale, J. *Afrique Noire: l'Ere Coloniale* (Editions Sociales, Paris, 1971); Eng. translation, *French Colonialism in Tropical Africa, 1900-1945* (New York, 1971).
——— *Afrique Noire: de la Décolonisation aux Indépendances.* (Editions Sociales, Paris, 1972).
——— *Afrique Noire, Géographie, Civilisation, Histoire* (Editions Sociales, Paris, 1973).
U.N.R.I.S.D. *Système Alimentaire et Société.* (Geneva).
U.N.U. Various publications, Tokyo.
Verneuil, P. 'Comment orienter l'investigation en milieu rural africain à partir de la relation entre échange inégal, développement inégal et transfert de valeur?', *AMIRA*, no. 22 (AFIRD, Paris, 1978).
——— 'Eléments pour une classification des non-salariés de la petite production en Afrique de l'Ouest et du Centre', *AMIRA*, no. 27 (AFIRD, Paris, 1979).
World Bank. *World Development Reports*, 1978, 1979, 1980.
World Confederation of Labour. *L'Agriculture: Manger, Vivre Mieux ou Faire de l'Argent?* (Brussels, June 1979).
Ziegler, J. *Main Basse sur l'Afrique* (Editions du Seuil, Paris, 1980).

of them must thus be reduced and limited to what is strictly necessary given the very limited resources of national economies and the impact of their cost on peasant resources. While imports of them must be limited, it is not, however, possible to do without inputs to improve productivity and considerably reduce extensive agriculture, which is a source of rapid deterioration of soils.

A solution must be found to the dilemma that consists in having recourse to the world market, whereas the goal is to delink from this market, and the need to modernize agriculture.

That is where the strategy of industrialization in Africa should be mainly concentrated, putting itself at the service of agricultural development.

A host of industries would be established throughout the rural areas of Africa which could meet agriculture's need for inputs of all sorts; the making of ploughs, carts and seeders, the production of fertilizers, insecticides, selected seeds, etc.

Here more research needs to be devised and carried out so that the products of these industries do indeed correspond to the particular characteristics and natural conditions of African agriculture. African peasants have had enough of the unfortunate experience with imported technology, that, in addition to its high cost, was sometimes ill-adapted to their working conditions and their natural environment. In order to limit the damage and reduce dependence, we think that the technical conception of these industries necessitates the collaboration of artisans, peasants and technicians trained in rural technical institutes, and the help of other underdeveloped countries that have registered striking successes with this sort of experience. We are thinking in particular of certain countries in Asia which were absolutely determined to free themselves from the tutelage of imperialism, to put an end to their domination and dependence.

Organization: We have already discussed at length the crucial need for powerful well-structured organizations of the peasantry as the prerequisite of rural development, and we must now reflect on the forms that these organizations might take to be able to put into effect the strategy that we have just outlined.

It might, for example, be possible to conceive a union of rural co-operatives which would have as its basic cell the whole village community united in a co-operative. The village co-operative would look after all the problems of the village community, even if the community were composed mainly of peasants in the vast majority of cases. What we mean is that it would be responsible also for solving the problems of herdsmen, fishermen, artisans, etc., in order to coordinate the activities of different groups of producers and create social harmony within the village community.

The co-operative will thus have a very extensive field of action. It will be responsible for the collection and marketing of the products of the whole village community. It will create distribution circuits and control the credit

Index